PROGRESS IN HPLC
Volume 3

PROGRESS IN HPLC
Volume 3

Flow through radioactivity detection in HPLC

Editors
H. Parvez, A. R. Reich, S. Lucas-Reich
and S. Parvez

///VSP///
Utrecht, The Netherlands,
1988

VSP BV (formerly VNU Science Press BV)
P.O. Box 346
3700 AH Zeist
The Netherlands

First published 1988

CIP-DATA KONINKLIJKE BIBLIOTHEEK, DEN HAAG

Flow through radioactivity detection / eds. H. Parvez . . .
[*et al.*]. — Utrecht : VSP. — Ill. — (Progress in HPLC;
vol. 3)
With ref.
ISBN 90-6764-114-6 bound
SISO 538.3 UDC 539.16.074
Subject heading: radioactivity; detection.

Printed in Great Britain by J. W. Arrowsmith Ltd., Bristol

CONTENTS

Preface

The progress of radioisotope techniques in the biomedical sciences has accelerated with the acceptance and growth of high pressure liquid chromatography. The emphasis is on speed, provided by the data processing capabilities of the sophisticated computer systems now integrated with laboratory instrumentation. Clean and simple automated techniques replace tedious and often obsolete methodologies to decrease the need for manpower. Many static measurements are being replaced by on-line dynamic measurement modes.

This volume in the Progress in HPLC series demonstrates a range of applications in which radio-high pressure liquid chromatography has proven to be not only useful but intrinsic to the success of an investigation. From plant metabolism to human clinical studies, the experiments described in this volume intend to form a guideline around which many new experiments can be built. A straightforward, simple set-up of HPLC with an on-line radioactivity detector as well as a totally automated RHPLC system illustrate the scope of this highly flexible modality. A brief chapter acquaints the reader with some significant steps in the development of radio-high pressure liquid chromatography and examins some of the problems associated with this technique.

Progress cannot be achieved without dedication. Progress in HPLC and in radioisotope methods has helped to advance biomedical research to its present state through the dedication of those scientists whose work constitutes this book. Dedication means inspiration, interest, concern, and most of all, commitment to a purpose.

We tried to encompass as wide a range of research carried out with radio-HPLC as was available to us for this volume. Certainly, there are many more experiments performed today in many more disciplines using this technique. We hope to learn about them and by publishing new information as it becomes available, share and advance the progress of radio-high pressure liquid chromatography.

We must express our appreciation and gratitude to the authors whose chapters are included in this book. We thank them for their willingness to share the results of their research and for the effort and time they have committed to the realization of this book.

Progress in HPLC, Vol. 3, pp. 1-10
Parvez et al. (Eds)

Radioactive flow detectors: History and theory

ANDREW R. REICH, SUSAN LUCAS-REICH
Radiomatic Instruments and Chemical Co. Inc., Tampa, Fl.
and
HASAN PARVEZ
Unité de Neuropharmacologie, Université de Paris,
Centre D'Orsay, 91405 Orsay Cedex, France

INTRODUCTION

In the early 1920's, it was established that small amounts of radiotracers used in biological processes do not alter or damage living organisms. At that time, radionuclides were already used in several therapeutic regiments, e.g., radium for the treatment of lupus and radiobismuth for the treatment of syphilis.

Diagnostic applications of radiotracers were first described for blood flow studies in humans using radium 228 (Blumgart and Yen, 1927). Radioactive indicators used in plants were described as early as 1923 (Hevesy,1923), and some years later, Hevesy began to experiment with the radioactive isotope of hydrogen for measuring the water content of the human body. Radioactive isotopes of trace elements were used to study absorption rates (Hamilton, 1938), and iodine 128 to investigate thyroid function (Hertz et al., 1938; Hamilton and Soley, 1939).

The invention of the cyclotron and subsequent production of various radionuclides opened up a new era in the biological and medical sciences. In the 1930's, Hamilton said that the discovery of artificial radioactivity and the development of the cyclotron have given the biologist the most useful tool since the discovery of the microscope.

Two studies revolutionized the development of biological sciences: the use of radioactive phosphorus in rat metabolism studies (Chievitz and Hevesy, 1935), and the discovery of a constant dynamic equilibrium process in a living organism (Schoenheimer, 1942).

In 1946, carbon 14 was produced for the first time in Oak Ridge, TN, making possible the development of new methodologies and applications of radionuclides in animal and human studies. Biological models began to provide the means to localize new biochemical substances of interest and follow their kinetic behavior as well.

In drug design, a substance under investigation must have physical or chemical affinity to a specific tissue to produce the desired action. The radionuclide used as carrier molecule for the parent compound must not affect these properties. When the compound is introduced into a living organism, it is absorbed and then distributed to various organs during its metabolic cycle. It might be transformed or eliminated partially or totally as it goes through a series of physical and chemical processes. The necessity to follow and study these processes created a great interest in radiotracer utilization and in the design of suitable instrumentation for their detection. As Colombetti stated, "two pillars support the entire edifice of nuclear biology and nuclear medicine: instrumentation and radiotracers."

Various methods are now available to substitute different atoms in a molecule with radionuclides. The most commonly used tracers are the radioactive isotopes of hydrogen , carbon, sulfur and phosphorus, which are the most abundant atoms in organic matter.

Radiotracer metabolism studies are carried out by *in vivo* and *in vitro* experiments. *In vivo* experiments contribute to the development of nuclear medicine in the process of diagnosis, the study of disease progression and the effectiveness of current therapy. *In vitro* turn over and binding studies are used to indicate the metabolic pathway of the administered substance and identify factors that affect its metabolic transformation processes. *In vitro* studies employ radioactive tracers either *in situ* (e.g., organ perfusion), or discreet specimens (tissues or cell homogenates, which can be disposed of when the experiment is finished).

Localization of the designed radiotracer makes it possible to find specific receptor sites. Studies of the chemical properties and physical characteristics of the receptor sites provide information on the kinetics of binding, the dissociation between the receptor and a particular radiotracer, and finally the parent compound under investigation.

The last 20 years have brought on an extensive increase in published material presenting the design and analysis of new labeled substances. According to Henry Wagner "an understanding of regional functions, i.e., organ function, cellular function, membrane function, and intracellular function will be more useful to the development of new labeled drugs than the idea of a magic bullet or any special attraction of an abnormal focus."

To realize the importance of radioactive tracers in nuclear sciences, new instrumentation must constantly be developed. In the last decade, tremendous progress has been achieved with the combined techniques of high pressure liquid chromatography and beta or gamma emitting radionuclide detection.

RADIOACTIVITY DETECTION FOR HIGH PRESSURE LIQUID CHROMATOGRAPHY

Progress in analytical chemistry during the last decade made high pressure liquid chromatography (HPLC) the prevalent separation technique. To analyze for radioactivity, samples from HPLC were fractionated and quantitated in liquid scintillation counters (LSC). Since this method was too slow and tedious compared to the speed and utility of HPLC, radiochromatography flow detectors were developed. The technique of radio-HPLC (RHPLC) evolved as the two scientific disciplines, high pressure liquid chromatography and radioactive flow detection were linked together into one logical analytical system. The early tentative attempts to develop a practical RHPLC detector were sometimes misdirected. These early detectors were liquid scintillation counters modified to do flow counting by people with limited chromatography backgrounds. The instruments were viewed as eccentric curiosities and could not be accepted by practicing chromatographers.

The earliest known commercial flow detectors were single sample LSC's modified to accept an anthracene filled vial (Schram and Lombaert, 1962; Albert et al., 1967). The inlet and outlet lines were sealed against light leaking into the counting chamber. The counter was set into a repeat count mode and by plotting the stream of numbers that resulted, a radiochromatogram was produced.

This attempt, unfortunately, set the tone of further developments. The thinking of both the users and developers was restricted to certain areas for the next decade. It was thought that the sensitivity of the detection was so low that the technique was only useful for samples with high absolute activity (Tkachuk, 1962; Schutte, 1972), that the use of solid scintillators was mandatory; and the results thereby obtained were, by necessity, unreliable. It took ten years for the first modern liquid-cell based system to change that undeservedly bad reputation (Schram, 1970; Reeve and Crozier, 1977; Snyder and Kirkland, 1979). Today, the discerning user can rely on the RHPLC detector to produce answers as fast and as reliably as the rest of the HPLC system. The results in units of radioactivity are now consistent with other radionuclide measurements and are reproducible within statistical limits.

PRINCIPLES OF RHPLC

To understand radio-HPLC, three basic concepts need to be discussed:

1. Normalization of results
2. Calculation of minimum detectable activity
3. Liquid cell vs. solid cell technique

Results normalization

In every analytical technique the most important criterion is the reproducibility of the results. This requirement must not only be true within the same experiment but must also be true external to the test. Results obtained by one method must be correlated to results from different experiments. Only in this way can a researcher be sure of the validity of the test used.

Early attempts to quantitate results from radioactive flow detectors failed in this respect. The results were reported as accumulated counts to emulate the fraction collecting with LSC counting technique. The presentation had little correlation with the results obtained by actually fractionating and counting the same sample in a liquid scintillation counter. Not understanding the nature of the phenomenon, exotic explanations were invented to rationalize the inconsistency of the results. To explain the differences between the two methods, the two detection efficiencies were correlated resulting in a new expression: ''flow counting efficiency''. It took several years to develop the theory of flow counting and postulate the mathematics behind the technique, thus eliminating such aberrant explanations.

Possibly the most important factor for misunderstanding RHPLC stems from the fact that the measure of radioactivity is different from any other chromatographic measurement. Contrary to UV transmission, IR absorption, or the angle of dispersion which are constant properties of the sample and can be measured independently of time, the measurement and quantitation of radioactivity are related to time. Both traditional and modern measures are based on time: one Curie is equivalent to the 2.22×10^{12} disintegration *per minute*, and one Bequerel equals one disintegration *per second*. While chromatographic measurements other than those of radioactivity, in quantity and accuracy are strictly dependent on detector sensitivity, time becomes a factor for radioactive flow detection. Time can significantly change the reproducibility and accuracy of results if it is insufficient or uncompensated for. The measurement of counting time in a flowing system is not obvious. The only way to visualize it is to view the effluent though the detector as a flow of discrete segments. These segments, as rungs of a ladder, follow each other through the flow cell and spend finite time in the cell. As infinitely small theoretical segments, their residence time in the flow cell depends on two parameters only : the size of the cell and the rate of the flow. By using the following equation, the residence time of each segment can be calculated:

$$R = \frac{V}{F}$$

R = Residence time in minutes
V = Cell volume in milliliters
F = Flow rate in ml/min

While in practice infinitely small theoretical segments are not measured, as long each segment is small compared to the size of the flow cell, the relationship is valid. Therefore, by taking frequent measurements in a flowing system, the true counting time and the true counts/time can be calculated with the flow equation. In a flowing system, using a 0.25 ml cell and a flow rate of 2 ml/min, the residence time is calculated to be 0.125 or 1/8th of a minute. As each segment is only counted for that time period, it follows, that the counts obtained from such a system, even if counted for one minute, do not represent counts per minute (cpm), but rather counts per 1/8th of a minute.

To normalize the count, that is to obtain true cpm results, the counting data must be multiplied by the inverse of the flow equation ratio; in this case by 8. It is now easy to see that in systems where the flow rate is higher, the counts accumulated are lower, so therefore, the apparent efficiency seems to be low. To compare runs using different size flow cells and different flow rates, the resulting counts must be normalized by using the flow equation as described.

Today, most modern digitally operated RHPLC systems use the correction to cpm on both the digital and analog data presented. The counting (or update) time per segment is either permanently set or given as a variable parameter to input data into the computer. Most systems use from two to six seconds as their "update" time. In systems where an analog output of the flow detector is coupled with "off the shelf" HPLC data processing software, this calculation is not available, therefore, the validity and reproducibility of the results are in jeopardy.

Indiscriminate use of the flow equation and flow corrected cpm, however, may cause a problem for the chromatographer. As seen above, the flow equation anticipates and corrects for counts resulting from segments of a flowing stream. Counts not resulting from flowing systems can not be treated with the flow equation as they will be grossly enlarged. Filling the flow cell with a known number of disintegrations per minute (dpm) of radioactivity (for the determination of counting efficiency) and counting it for a set amount of time (1 min) is a standard practice. The use of the flow equation in this case however, would result in erroneous results, since the activity is not in segments and not in a flowing mode. Consider this experimental situation:

The following conditions are given:

Cell size: 1 ml
Flow rate: 4 ml/min
Update time: 6 sec
Standard solution: 15000 dpm/ml
Counting efficiency: 50%

For a one minute counting period (ignoring statistical variations) the system will accumulate the following 10 six second counts:

750 750 750 750 750 750 750 750 750 750

If the flow equation is not used, the sum is 7500 cpm, which is the correct number. However, by using the equation, each segment will be subject to the inverse of the V/F calculation. In this case, each segment will be multiplied by 4, and in one minute the following 10 counts will be summed:

$750 \times 4 = 3000$

3000 3000 3000 3000 3000 3000 3000 3000 3000 3000

The resulting sum of 30,000 cpm is obviously an invalid number. The only way this number could be valid is if that during the one minute counting period, 4 ml of the standard solution passed through the cell, and the result was divided by the flow rate.

A more disturbing phenomenon occurs with the background subtraction parameter in the flowing system. The true background of a system is not the result of the flow, but is caused by cosmic radiation and electronic noise that are constantly present and independent of the flow. In order to obtain valid results, this "static" background must be eliminated before the flow equation is applied. If not, the real background is overstated by the multiple of the flow rate, as was shown with the 10 six second counts in the similar example above.

Other than these pitfalls, the flow equation has been shown to be a valid and effective method to normalize results. A comparison between chromatographic runs using different flow cells and different flow rates can be made with an expected correlation. More significantly, in recovery studies, the preinjected dose can be measured in a liquid scintillation counter and the resulting cpm or dpm value can be compared to the results from the RHPLC without exotic correlation coefficients or other factors. The use and understanding of this flow equation moved RHPLC into the realm of accepted analytical methodology.

Minimum detectable activity (MDA)

Being an experimental science, HPLC parameters in any given run are determined by the time tested method of trial and error. So, for many years, the minimum detectable limit of activity in an RHPLC sample was determined by successive injections of diminishing activity in serially diluted standards of known retention time. For the sake of expediency, chromatographic conditions were generally set to make the standard elute in the shortest possible time (two to five minutes) so that the "real" experiment could be started with little delay. This practice has been successful in determining the MDA of the standard, but has not necessarily helped with the MDA of the compound under investigation.

A few flow detector manufacturers have determined and published MDA data, some footnoted by the statement that the published MDA is for compounds behaving normally under chromatographic conditions. Practicing chromatographers know that virtually all compounds behave abnormally or at least unexpectedly under chromatographic conditions making the disclaiming statement unacceptable. In fact, experimental data is available to demonstrate that the same compound under different chromatographic conditions may behave differently concerning its MDA. Compounds well resolved and eluting early in sharp well proportioned peaks can be detected in smaller quantities than late eluting, shallow and broad peaking compounds.

We have derived an equation to determine the MDA based on the the relative peak height and the presumption that the limit of detectability can be defined as twice the count rate of the background. The following formula where the MDA is in dpm, correlates well with experimental data and can be used in any application:

$$MDA = \frac{B \times W}{T \times E}$$

B is the background in cpm
W is the peakwidth in minutes
T is residence time in minutes as derived above (T=V/F)
E is efficiency percent over 100

While we do not intend to derive this formula here, its theory is based on the following principle: *The Minimum Detectable Activity* of a flow detector is not directly related to the total amount of activity in any given peak, but rather to the specific activity of any flow segment residing in the detector at any one time. In other words the sharper the peak, the easier it is to detect. If the specific activity (the peak) is twice the background of the detector, the peak is detectable; otherwise it is not. Therefore, even significant amounts of activities running through the detector in large volumes of eluent can lower the specific activity of the mixture significantly enough (thus creating a large elongated peak), to escape detection. By using this simple equation, one more mystery may be eliminated in the practice of RHPLC.

Liquid vs. solid cells

The final mystery of RHPLC is centered around the use of liquid vs. solid scintillator cells to detect the radioactivity in the effluent.

Most of the early work was done exclusively with anthracene filled tubes coiled inside LS counters. Later, as investigators learned that the particle size of the anthracene had an inverse effect on the counting efficiency, and many of the converted liquid scintillation counters were being ruined by floods of solvent because of blocked anthracene coils, the race for a better scintillator was on. The problem with anthracene was that as the particle size diminished, the soft organic crystals packed together, forming a plug impenetrable by the flowing effluent. The smallest particle size still servicable was about 500 microns, and counting efficiencies for carbon 14 in the area of 55% to 60% while tritium efficiencies in the 0.01% to 0.05% were recorded. A breaktrough of sorts was made when Laboratorium Berthold introduced a solid scintillator filled cell containing a "mystery" compound that raised the tritium counting efficiency by tenfold. The mystery compound turned out to be lithium glass, a by-product of high energy nuclear physics research. Utilizing the incompressibility of the glass, a filling of a much reduced particle size can be produced and still allow the liquid to flow through the cell. Other scintillating inorganic crystals of similar rigidity are in use today which can raise the tritium counting efficiency of the solid scintillator cell into the low single digit area. Higher counting efficiencies for tritium however, did not come without problems. The better the counting efficiency of the cells, the more they became prone to contamination. While a liquid cell may last for the life of the detector, solid cells must be constantly changed, washed, repacked and otherwise tended to. Results again become questionable due to artificially high background subtractions required besides other problems. The apparent simplicity of the solid cell technique is, therefore, misleading.

The theory of radioactive flow detection can shed some light onto the source of this problem. The energy of the particles released during disintegration is very low. The physical value of a keV may not mean very much for the practicing chromatographer, but a comparison may. The average energy of a beta particle released during the disintegration of a tritium atom is 15.4 keV. This is such a low energy level that the beta particle cannot penetrate more than one millimeter of air. Needless to say, since water is much more dense than air, the penetration of the beta particle from tritium is less then one micron in any liquid medium.

Counting tritium in a liquid scintillation medium, such as xylene, has the advantage that while the beta particle is absorbed by the first organic solvent molecule it encounters, the energy absorbed by this molecule is then released in the form of an electron. The electron then is absorbed into the next organic solvent molecule, and the process is repeated, like billiard balls transmitting motion, until a scintillator molecule is encountered. At that point the absorbed electron does not emerge, but is transformed into an other form of energy, and released in the form of a photon. The reason this process can take place is the loose electron cloud surrounding all efficient LSC solvents.

The energy absorbed into water, or into most commonly used HPLC eluent systems disappears and is not able to produce scintillation: therefore, the counting efficiency of the system suffers. For a solid scintillator particle to have high efficiency, it must not have dimensions larger than that of the molecule containing the radioactive tritium. They need to be in intimate contact for the beta particle to reach the scintillator. By reducing the size of the solid material in the counting cell, this condition can be approached, thus, higher counting efficiencies can be achieved. Lowering the particle size, in terms of chromatography, unfortunately connotates another phenomenon. Both the absorption and the adsorption of a compound on a solid surface are favorably influenced by diminishing particle size. This phenomenon is used in chromatography: for better separation, increasingly smaller sized support is used. When the size of the detector cell packing approaches the size of the column packing, suddenly, the chromatographer has two columns. One, (his HPLC column) he can control, the other,(the detector) he can not. Therefore, effects of unpredictable character may occur right in the detector cell. Compounds, radioactive or not, may adsorb on the finely granulated solid support, staying in the cell for periods of time dependent upon their individual characteristics. This will affect the results in two different ways. One is predictable, annoying, and if not preventable is at least visible. The other, being unobservable is more serious then the former. If a radioactively labeled compound is adsorbed on the scintillator, the detector becomes *contaminated*. This contamination may be short term, in which case the observed radioactive peak becomes broadened. The effective flow rate of the compound through the cell becomes less than that of the eluent; therefore, the flow equation becomes invalid. The results are erroneously high, and the chromatographer's effort to optimize conditions are wasted while trying to remedy one problem when the problem is somewhere else. In case of the long term contamination, the only remedy is changing the cell and removing the adsorbed compound.

Adsorption of nonradioactive material on the scintillator is probably even more troublesome, as the user does not know when it occurs, and what to do about it. The effect of nonradioactive contamination on the scintillator is the loss of counting efficiency, since the counting efficiency is dependent on the availability of the surface of the scintillator. Just as with radioactive contamination, short or long term, nonradioactive contamination can occur resulting in bad results for short or long term without warning.

How should solid scintillators be used with success? Is there a reason to do so?

The recommendation of a manufacturer to precoat the scintillator with cold contaminating material obviously cannot be accepted. It appears that for the detection of low energy nuclides, there may be more problem than benefit derived from the use of solid scintillators. There are a select few cases where there may be an argument for their use: The investigator knows the composition of the sample, therefore, the question is quantitation; the investigator has evidence that none of the compounds in the sample mixture have an affinity to the scintillator; the level of activity of the sample is high enough to overcome the low counting efficiency of the solid scintillator. As the energy level of the radioactive compound increases, the problems mentioned above become less innocuous. If contaminated with carbon 14 or phosphorus 32, there will be no question that the cell will have to be changed. Peak broadening with these materials will cause as much trouble as with tritium, but the coating of the active surface will not significantly lower the efficiency of the cell.

The best advice still is to use solid scintillator cells *only* with known compounds; in situations where controls are available and comparison with nonradioactive runs can be made easily. For the analysis of unknown mixtures, such as metabolites, use of the liquid cell scintillation technique is the only sensible method. It may be required to dispose of some spent scintillator, but the validity of the results, the sensitivity of the analysis and the unquestionable reproducibility of the experiment must be worth the extra effort. After all, if the results are not important enough why run the experiment?

CONCLUSION

During its first two decades, radio high pressure liquid chromatography became a routine and accepted analytical tool. The speed of HPLC coupled with on-line radioactivity counting made this the technique of choice for many researchers. In the following pages many applications are described; some serve as a testimony to the method; some serve corollary to other more important aspects. In all cases, whether documented or not, one can assume that the research presented in this book could not have been done as quickly as it has been, without RHPLC. For all the controversy, all the problems solved and still waiting for resolution, one aspect is certain: RHPLC has become a very important method in the struggle to understand the structure and nature of biochemical systems, playing a part to help in the fight against disease.

REFERENCES

Albert, S,N., Zekas, E., and MackTimby, R. (1967) Anthracene cells for counting weak-beta-emitting nuclides (Sulfur-35) in liquid samples. *J. Nucl. Med.* 8, 822.

Blumgart, H.L., and Yens, O.C. (1927) Studies on velocity of blood flow, Method utilized. *J. Clin. Invest.* 4, 1.

Chievitz, O., and Hevesy, G. (1935) Radioactive indicators in the study of phosphorus metabolism in rats. *Nature,* (London) 136, 754.

Hamilton, J.G., and Alles, G.A. (1938) Physiological action of natural and artificial radioactivity. *Am. J. Physiol.* 124, 667.

Hamilton, J.G. and Soley, M.H. (1939) Studies of iodine metabolism by use of new radioactive isotope of iodine. *A. J. Pysiol.* 127, 557.

Hertz, S., Roberts, A., and Evans, R.D. (1938) Radioactive iodine as indicator in study of thyroid physiology. *Proc. Soc. Exp. Biol. Med.* 38, 510.

Hevesy,G. (1923) Absorption and translocation of lead by plants. Contribution to the application of the method of radioactive indicators in the investigation of the change of substance in plants. *Biochem. J.* 17, 432.

Reeve, D.R., and Crozier, A. (1977) Radioactivity monitor for high performance liquid chromatography. *J.Chromatogr.* 37, 271.

Schoenheimer, R. (1942) In: *The Dynamic State of Body Constituents,* Harvard University Press, Cambridge.

Schram, E., and Lombaert, R. (1962) Determination of tritium and carbon-14 in aqueous solutions with anthracene powder. *Anal. Biochem.* 3, 68.

Schram, E. (1970) Flow-monitoring of aqueous solutions containing weak B-emitters. In: *The Current Status of Liquid Scintillation Counting,* E.D. Bransome, Jr. M.D.(ed.) Grune and Stratton, New York, pp. 95-109.

Schutte, L. (1972) Continuous detection of radioactive effluents in liquid chromatography by heterogeneous or homogeneous scintillation counting. *J. Chromatogr.* 72, 303.

Snyder, L.R. and Kirkland, J.J. (1979) In: *Introduction to Modern Liquid Chromatography,* John Wiley & Sons, Inc. New York, pp. 158-161.

Wagner, H. Jr. (1977) Nuclear medicine in motion. *J. Nucl. Med.* 18, (1), 2.

Progress in HPLC, Vol. 3, pp. 11-26
Parvez et al. (Eds)

Use of radiotracer techniques and HPLC with flow scintillation detection in the analysis of fatty acids and eicosanoids

DALE L. BIRKLE, HAYDEE E. P. BAZAN and NICOLAS G. BAZAN
Louisiana State University Eye Center, New Orleans, Louisiana, USA

SUMMARY
Techniques for the HPLC of fatty acids and oxygenated products of arachidonic acid (eicosanoids) are described. The chemistry and biological significance of these compounds are reviewed briefly. The advantages, disadvantages and special applications of radiotracer methodology and HPLC are discussed. Several HPLC systems for the separation and quantitation of radiolabeled fatty acids and eicosanoids are described. These include reverse phase and straight phase HPLC systems. Special problems in the preparation of lipid samples for HPLC are discussed, and several aspects of troubleshooting an HPLC with flow scintillation detection are described.

1. CHEMISTRY OF FATTY ACIDS AND EICOSANOIDS

1.1 Fatty acids

Fatty acids are straight-chain hydrocarbons containing a carboxylic acid function at their α-end. They range in chain length from 14 to 24 carbons and are saturated or unsaturated, containing one or more carbon-carbon double bonds. Naturally occurring fatty acid alkenyl groups have a methylene-interrupted *cis* configuration (Van Golde and Van den Bergh, 1977). Table 1 shows a list of biologically significant fatty acids with their common and systematic names. The system of nomenclature for fatty acids is "chain length: number of double bonds (position of the first double bond)." For example, arachidonic acid's systematic name is eicosatetraenoic acid and is abbreviated 20:4 (n-6) because it contains 20 carbons (eicosa...), and four double bonds (tetraenoic) beginning at the sixth carbon (n-6) from the omega (methyl) end. There are two major families of polyunsaturated fatty acids: the linoleic acid family with the n-6 structure, and the linolenic acid family with the n-3 structure. In mammals, both linoleic and linolenic acids are essential fatty acids and must be provided by the diet

Table 1. Fatty acids of biological interest

Common Name	*Systematic Name*	*Nomenclature*
Myristic acid	n - tetradecanoic	14:0
Palmitic acid	n - hexadecanoic	16:0
Stearic acid	n - octadecanoic	18:0
Palmitoleic acid	9 - hexadecenoic	16:1
Oleic acid	9 - octadecenoic	18:1 (n-9)
Linoleic acid	9,12 - octadecadienoic	18:2 (n-6)
α-Linolenic acid	9,12,15 - octadecatrienoic	18:3 (n-3)
γ-Linolenic acid	6,9,12 - octadecatrienoic	18:3 (n-6)
Arachidonic acid	5,8,11,14 - eicosatetraenoic	20:4 (n-6)
Docosapentaenoic acid	4,7,10,13,16 - docosapentaenoic	22:5 (n-6)
Cervonic or docosahexaenoic acid	4,7,10,13,16,19-docosahexaenoic	22:6 (n-3)

(Van Golde and Van den Bergh, 1977). By means of elongation and desaturation pathways, occurring mainly in the liver (Van Golde and Van den Bergh, 1977), linoleic and linolenic acids are converted to the polyunsaturated fatty acids, e.g., docosahexaenoic acid (22:6, n-3) and arachidonic acid.

1.2 Eicosanoids

The eicosanoids are oxygenated products of arachidonic acid (Corey et al., 1980). Prostaglandins (PGs) are synthesized by cyclooxygenase; hydroxyeicosatetraenoic acids (HETEs) and leukotrienes (LTs) are synthesized by lipoxygenase (for reviews see Hammarstrom, 1983; Lands, 1979). PGs are derivatives of prostanoic acid, containing a five membered ring with various functional groups attached, and two side chains. The α-side chain has the carboxylic acid function, and the β-side chain terminates in a methyl group (Fig. 1). The nomenclature is based on the substitutions in the ring (Fig. 2). Lipoxygenase reaction products, e.g., the HETEs, can occur as

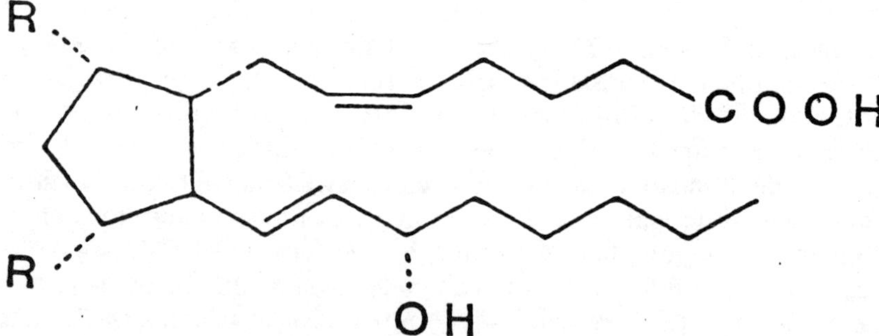

Figure 1. Basic structure of the prostaglandins. Carbon-1 is the carboxylic function, located on the α-side chain. The 2β-side chain terminates at carbon-20, the methyl group. One, two or three double bonds are present, depending on the unsaturation of the fatty acid precursor. There can be various functional groups on the ring. Structures of individual prostaglandins are shown in Figure 2.

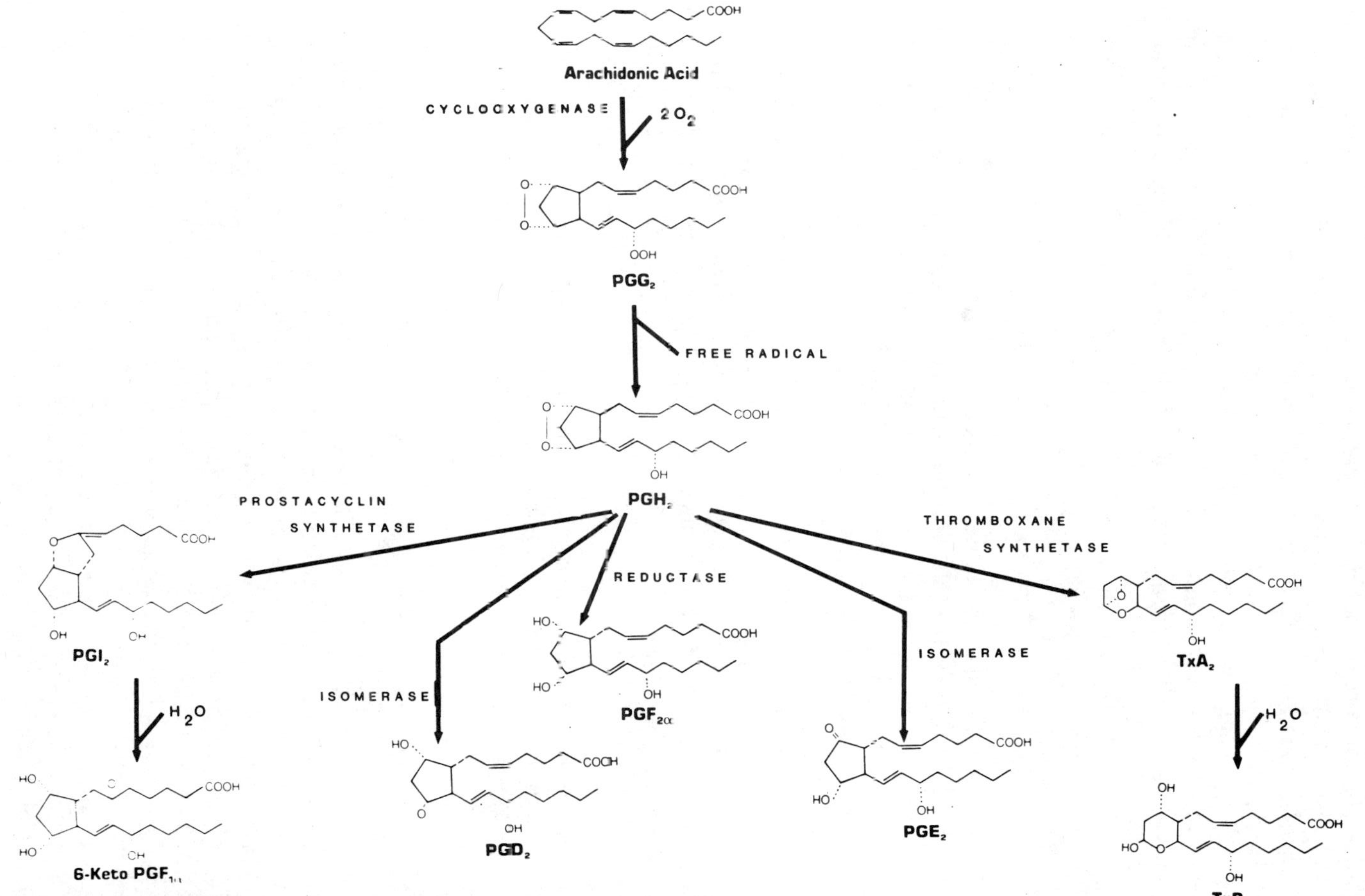

Figure 2. Synthesis of cyclooxygenase reaction products of arachidonic acid.

various isomers depending upon where addition of oxygen takes place. The immediate precursors of HETEs are hydroperoxy derivatives. Further metabolism of HETEs by addition of peptide moieties results in the synthesis of LTs (Fig. 3). Eicosanoids can also be synthesized from other fatty acids, such as eicosatrienoic acid (20:3, n-6) and eicosapentaenoic acid (20:5, n-6). These metabolites are denoted by the subscript, which indicates the number of double bonds. Hence, PGs from arachidonic acid are the 2-series, e.g., PGE_2. PGE_1 is a product of 20:3 (n-6) and PGE_3 is a product of 20:5 (n-6).

2. BIOLOGICAL SIGNIFICANCE

2.1 Fatty acids

In mammalian tissues, e.g., brain, fatty acids are present in their free form in very small quantities (Bazan, 1970). Mainly, they are found esterified to glycerol as part of glycerolipids. In general, phospholipids contain a saturated fatty acid esterified to carbon-1, a polyunsaturated fatty acid esterified to carbon-2, and a phosphate-base group esterified to carbon-3 (Van Golde and Van den Bergh, 1977). The chain length and number of unsaturations in these lipids influence the fluidity of a membrane, while the phosphate-base groups influence the polarity or ionic charge of the membrane (Cullis et al., 1980). Therefore, fatty acids play a crucial role in determining the lipidic environment of the cell membrane, both in the plasma membrane with its receptor/enzyme complexes and in the interior membranes, e.g., the smooth and rough endoplasmic reticulum. Fatty acids are also found as part of proteins (Bazan et al., 1985). To increase the solubility of fatty acids in aqueous solutions, they are carried in serum bound noncovalently to serum albumin and other carrier proteins; they also occur as integral parts of lipoproteins and proteolipids.

2.2 Eicosanoids

PGs, HETEs and LTs have potent actions in many, if not all, cells and tissues (for reviews see Goodwin and Ceuppens, 1983; Wolfe, 1982; Hammarstrom, 1983). They act as mediators and modulators of the inflammatory response by influencing chemotaxis, phagocytosis, and lysosomal enzyme release. They have potent vascular effects, causing vasodilation, vasoconstriction, changes in capillary permeability, and platelet aggregation. There is increasing evidence that eicosanoids are modulators of thermoregulation, gastric acid secretion, and various aspects of the reproductive cycle. LTs are extremely potent bronchoconstrictors and are probably the mediators of the asthma of anaphylaxis. Generally these compounds are produced in extremely small quantities in any given tissue. Therefore the measurement of eicosanoids in biological systems presents an important challenge to the analytical biochemist.

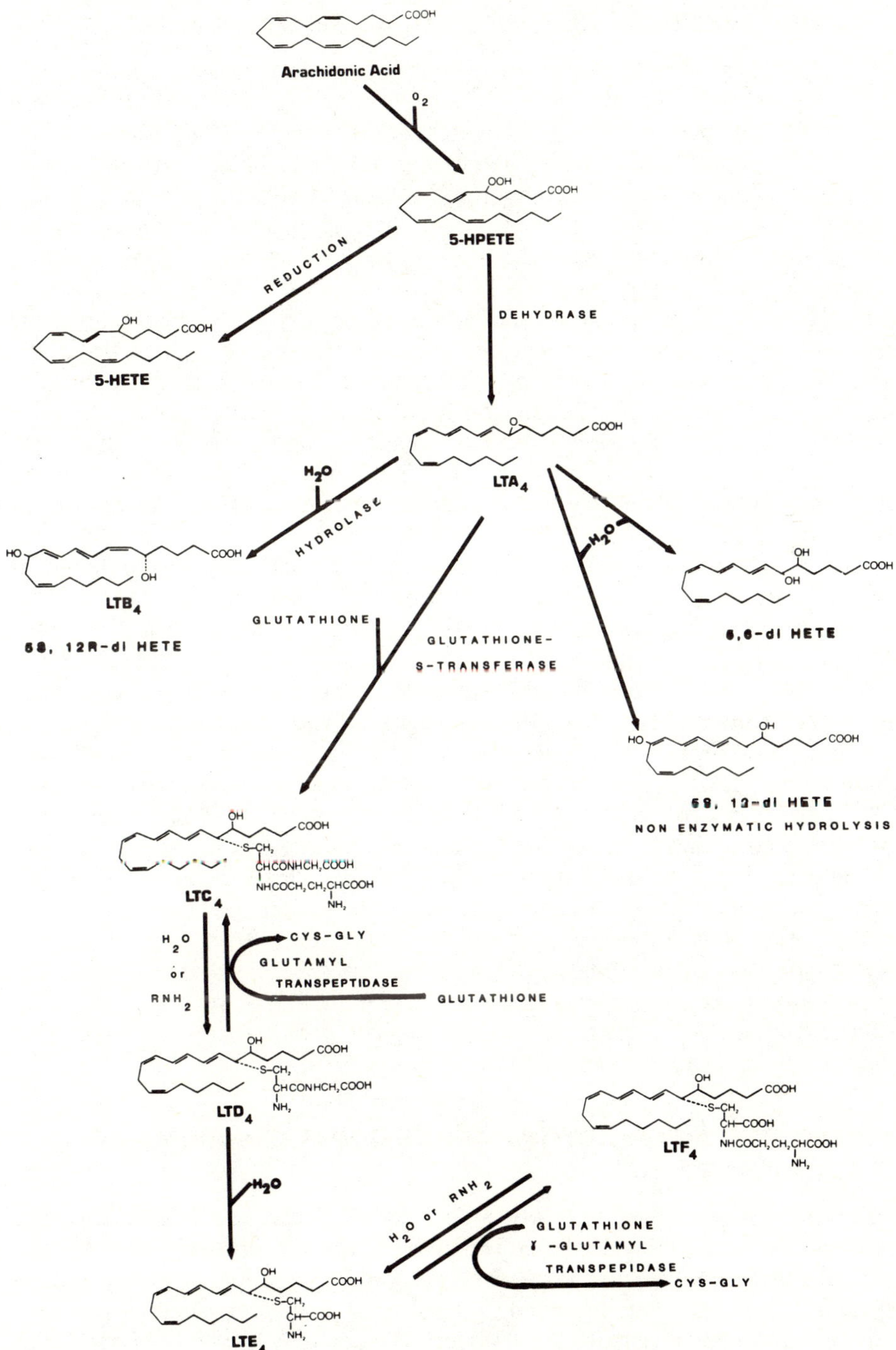

Figure 3. Synthesis of lipoxygenase reaction products of arachidonic acid.

3. HIGH PERFORMANCE LIQUID CHROMATOGRAPHY (HPLC) AS OPPOSED TO OTHER ANALYTICAL TECHNIQUES

HPLC offers several advantages over analysis of fatty acids and eicosanoids by the alternative techniques, gas liquid chromatography (GLC) and thin layer chromatography (TLC). GLC requires the synthesis of volatile derivatives, e.g., methylation of carboxylic acid functions, methoxylation of keto groups, and conversion of free hydroxyl groups to trimethylsilyl derivatives. These procedures add to preparation time, and provide opportunities for sample loss. The high temperatures used in GLC can cause decomposition, which is a major problem in the analysis of eicosanoids. GLC is an extremely powerful technique for the separation of fatty acids, resolving on the basis of boiling point, and therefore, capable of separating *cis-trans* isomers. However, it cannot resolve a mixture of PGs or HETEs, which in derivatized form are positional isomers that do not differ greatly in their boiling points or polarities. HPLC separates underivatized compounds mainly on the basis of polarity, and can perform a difficult separation like the resolution of PGE_2 (9-keto-11-hydroxy), and PGD_2 (9-hydroxy-11-keto), or 12-HETE, 5-HETE, and 15-HETE.

HPLC is a nondestructive technique, and therefore provides a convenient method for the purification and collection of fatty acids and eicosanoids. Because of the greater number of theoretical plates in a HPLC column, HPLC has a much greater resolving power than TLC has. Also, recovery of compounds by collection of HPLC eluent is simpler and more efficient than recovery from a TLC plate, which requires scraping the silica gel and eluting the compounds. Quantification of unlabeled fatty acids and eicosanoids eluted from a HPLC column can be done by radioimmunoassay (RIA), or gas chromatography-mass spectrometry (GC-MS). Quantitation of labeled compounds can be done most conveniently by on-line flow scintillation detection, thus avoiding the tedious process of collection of fractions, taking aliquots and counting in a liquid scintillation counter. Because in many cases interference and cross-reactivity produce uncertainty in the quantitation of eicosanoids by RIA, HPLC is a very suitable method for separation of the components in a sample prior to RIA. Similarly, purification of the sample by HPLC is necessary prior to quantitative or qualitative analysis of eicosanoids by GC-MS.

4. HPLC SET-UP FOR AUTOMATED ANALYSIS OF FATTY ACIDS AND EICOSANOIDS

Analysis of eicosanoids and fatty acids by HPLC can be a time consuming process, because each sample may require an hour or more of analysis time. This disadvantage of HPLC can be overcome by the use of an automated system, which allows unattended 24-hour operation. Our laboratory has set up such a system consisting of a Waters Associates (Milford, MA) Model 840 data station, a WISP (Waters, Model 710) sample processor, solvent delivery pumps, a Z-module radial compression unit for use with radial compression columns, a variable wavelength ultraviolet (UV) detector, and a Radiomatic FLO-ONE/Beta flow scintillation detector. Event signals generated by the 840 control the start and stop of the scintillation counting program and the pump for the scintillation fluid. The 840 is capable of storage of UV data, and the FLO-ONE/Beta can store radioactivity data, thus enabling the reprocessing of HPLC runs. The use of large solvent reservoirs (4 L for HPLC solvents, and 20 L for scintillation fluid) allows unattended operation for up to 24 hours, using an HPLC flow rate of 3 ml/min and a scintillation fluid flow rate of 6 ml/min.

5. RADIOTRACER TECHNIQUES FOR LIPID BIOCHEMISTRY

The factors that control the basal and stimulated metabolism of lipids in mammalian cells are poorly understood. This is an important area of research, particularly in light of new experimental evidence pointing to critical functional roles of lipid-derived compounds in various tissues (for reviews see Birkle and Bazan, 1985, 1986; Horrocks et al., 1986; Bleasdale et al., 1985). As described briefly in section 2, fatty acids and eicosanoids appear to have a number of physiological and pathophysiological roles.

One of the major obstacles in the study of lipid metabolism is the quantitation of various components of the biosynthetic and catabolic routes. The use of radiolabeled precursors has simplified this task considerably. Introduction of high specific activity radiotracers to a tissue allows labeling of various metabolic pools and subsequent quantitation by scintillation counting. These techniques require several assumptions, which should be verified if possible. They include: (1) adequate mixing of radiolabel with endogenous pools, (2) no effect of addition of radiolabel, (3) equal recovery of radiotracer and endogenous components.

We have used radiotracer techniques in several experimental protocols to study the metabolism of fatty acids, glycerolipids and eicosanoids in ocular tissues (Bazan et al., 1985; Birkle and Bazan, 1984a) and the central nervous system (Pediconi et al., 1982, 1983; Reddy and Bazan, 1983). A very useful protocol is the prelabeling method. Radiolabeled lipid precursors (e.g., fatty acid or glycerol) are introduced *in vivo* into brain by intraventricular injection, or into the eye by intravitreal injection. Preliminary time course data is obtained to determine maximum, steady-state incorporation of the precursor into the lipid pools of interest. Because of loss of radiolabeled precursor into the blood and peripheral tissues (Birkle and Bazan, 1984a), relatively large amounts must be injected; about 1 μCi per eye and 2 μCi per rat brain for fatty acids. However, high specific activity radiotracers represent very little mass, so effects are minimal. This point is very important in the case of fatty acids, which have detergent and other effects on cell membranes (Creutz, 1981; Rhoads et al., 1983). After the appropriate prelabeling time, stimuli can be applied *in vivo* (e.g., electroconvulsive shock, flicker light, cryogenic injury, carotid occlusion) and the effects on various lipids can be assessed. Alternatively or additionally, prelabeled tissues can be removed and subjected to experimental manipulation *in vitro*.

The prelabeling paradigm is advantageous because it allows mixing and incorporation of the radiotracer in a more physiological manner. However, this method is not always practical, for example, when the tissue source is a large animal like cow, or when postmortem tissues are used. In these situations, *in vitro* incubation of the tissues with radiotracer is used. A pulse-chase design can approximate the prelabeling paradigm, in which tracer is added, incubated, the excess washed out, and the stimulus applied.

A third use of radiotracer is in the measurement of endogenous lipids. This is particularly useful in the quantitation of eicosanoids. The tissue sample is obtained and known quantities of very high specific activity standards are added. The presence of the standards allows detection of the compounds of interest by scintillation counting and also serves as a control for the recovery of compounds through the extraction and isolation procedures. In the case of eicosanoids, addition of a combined deuterated/tritiated internal standard allows quantitation of endogenous levels by gas chromatography-mass spectrometry (Green et al., 1973). Endogenous phospholipids can be quantitated by phosphorus estimation (Rouser et al., 1970) and fatty acids can be quantitated by GLC (Bazan, 1970).

6. HPLC ANALYSIS OF FATTY ACIDS

For studies on turnover or metabolic conversion of fatty acids, reverse phase HPLC using a radioactive detector is very useful. This method has advantages over GLC using flame ionization detection, mainly because it allows injection of larger samples and detects radioactivity more accurately. Fatty acids can be analyzed either as free fatty acids or as methyl esters.

6.1 Preparation of the sample

Once the lipid extract has been obtained (Folch et al., 1957; Saunders and Horrocks, 1984), total fatty acids can be prepared by hydrolysis of glycerolipids. For preparation of free fatty acids, alkaline hydrolysis is used (Aveldano and Horrocks, 1983):

—Prepare a solution 0.5 N NaOH in methanol: water (9:1 v/v)
—Dry samples under N_2
—Add 1 ml of 0.5 N NaOH solution
—Put samples under argon
—Incubate overnight in a water bath at room temperature
—Acidify to pH 3 with 88% formic acid
—Extract twice with 3 ml hexane and twice with 3 ml ether, combining the organic phases
—Wash the organic phase twice with 3 ml of distilled water to remove acid
—Dry the organic phase under N_2
—Resuspend in a small volume of acetonitrile
—Count an aliquot of the sample to determine total radioactivity

This preparation can be analyzed by HPLC as free fatty acids, or, methyl esters can be made by reaction with ethereal diazomethane.

Total fatty acid methyl esters can be prepared directly from the lipid extract by methanolysis in boron trifluoride methanol, according to the procedure of Morrison and Smith (1964). In addition to analysis of total fatty acids, the fatty acid content and composition of individual glycerolipids can be determined by isolation of the lipids of interest prior to methanolysis or hydrolysis. Also, free fatty acid levels can be assessed by removal of the esterified fraction (glycerolipids) prior to methanolysis and HPLC.

6.2 Separation of free fatty acids

Labeled free fatty acids can be separated efficiently and their radioactivity detected using a reverse phase HPLC column, gradient elution with acetonitrile and aqueous phosphoric acid (pH 2.3), and flow scintillation detection (Aveldano et al., 1983). There are several factors to consider in the analysis of fatty acids by HPLC. There are two types of reverse phase columns that have been employed successfully in our hands for the separation of polyunsaturated fatty acids: the phenyl (C8) and octyldecyl (C18) columns. Because of the higher polarity of the C8 column, there is less avid retention of highly nonpolar fatty acids (e.g., palmitic acid, 16:0), thus offering an advantage in terms of analysis time over the C18 column. However, if saturated fatty acids are the compounds of interest, the C18 column offers better resolution of various chain lengths as compared to the phenyl column. The order of elution of fatty acids in a complex mixture will vary with the acetonitrile concentration, so it is necessary to change the solvent strength to afford separation of the fatty acids of interest. In general, the column selection and solvent strength must be tailored to separate optimally the particular fatty acids needed. Another factor to consider involves the injection of the sample. Samples are usually dissolved in the initial conditions solvents, which can contain as much as 70% water. This severely limits the solubility of saturated fatty acids. We have found that injection in 100% acetonitrile does not have deleterious effects on the separation, but can shift retention times as much as 3-4 minutes. Therefore, it is critically important to establish the retention times of fatty acid standards and the sample, under the same conditions, including injection parameters.

The separation of two fatty acids having the same chain length and different degrees of unsaturation is illustrated in Figure 4. Glycerolipids extracted from rat retina were subjected to alkaline hydrolysis, as described above, and the free fatty acids were separated on a C8 μBondapak (Waters Assoc.) column (8 mm x 10 cm, 10 μ) using an acetonitrile concentration of 40% in aqueous phosphoric acid (pH 2.5) for the first 40 minutes, with a flow rate of 3 ml/min, a linear increase to 60% acetonitrile over 10 minutes and another increase at 60 minutes to 100% acetonitrile to strip the column. This system was optimized for the separation of long-chain polyunsaturated fatty acids like docosahexaenoic, docosapentaenoic, arachidonic, and eicosapentaenoic acids. Saturated fatty acids eluted as a single peak during the 100%

acetonitrile strip. Carbon 14 was detected using a Radiomatic flow scintillation detector (Model HS) that was programmed to calculate total dpm with a counting efficiency of 56%. FLO-SCINT II liquid scintillation cocktail (Radiomatic) was used in a ratio of 3:1 (v/v) cocktail to HPLC eluent.

6.3 Separation of fatty acid methyl esters

Fatty acid methyl esters (FAME) are separated by HPLC using systems similar to those used for free fatty acids (Aveldano et al., 1983). Acidified solvents are not required because the carboxyl functional group has been methylated prior to analysis. In many cases, an acidified solvent system for free fatty acids can be used for separation of the same types of FAME, with the only modification being omission of the acid. Fig. 5 shows a separation of several radiolabeled FAME standards. The sample was injected into a C18 μBondapak radial compression column (8 mm x 10 cm, 10 μ). The solvent system was acetonitrile:water (70:30) at 2 ml/min for 55 minutes, followed by a linear increase to 100% acetonitrile over 10 minutes and a hold at final conditions for an additional 15 minutes. Radioactivity was detected with a Radiomatic FLO-ONE/Beta system, programmed to count carbon 14 with a counting efficiency of 70%. Ready-Solv EP (Beckman)

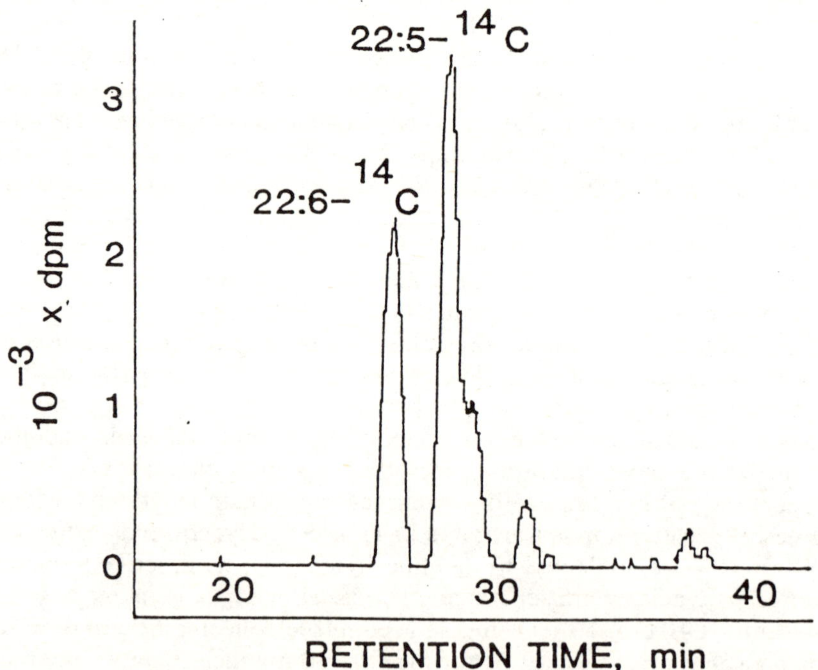

Figure 4. Separation of radiolabeled unsaturated free fatty acids by reverse phase HPLC. Radioactivity was monitored by a flow scintillation detector FLO-ONE (Model HS, Radiomatic, Tampa, FL). Shown is the conversion of docosapentaenoic acid (22:5) to docosapentaenoic acid (22:6) by the rat retina, *in vivo*. The samples (lipid extract from tissue equivalent to 150 μg of protein) were injected in 25 μl of acetonitrile.

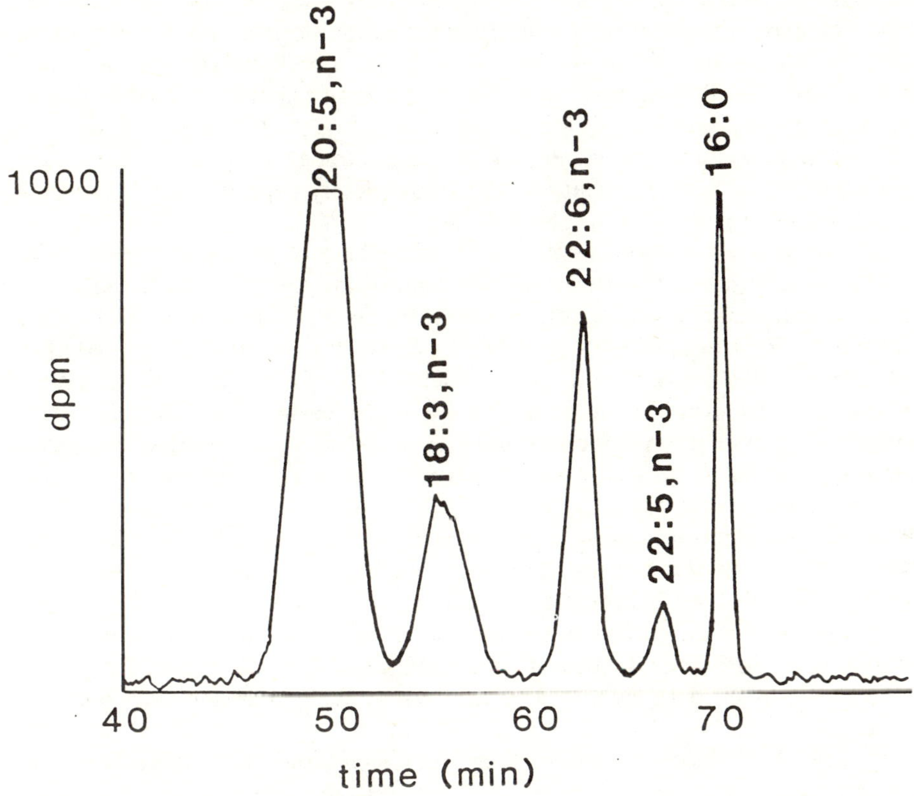

Figure 5. Separation of radiolabeled fatty acid methyl ester standards.

liquid scintillation cocktail was used in a ratio of 3:1 (v/v) cocktail to HPLC eluent.

7. HPLC ANALYSIS OF EICOSANOIDS

Experiments using labeled arachidonic acid as precursor of prostaglandins and lipoxygenase products have been employed in many tissues (Jacschik et al., 1978; Birkle and Bazan, 1984a, 1984b; Bazan et al., 1985). These metabolites can be separated by a variety of methods using reverse phase or straight phase HPLC (for review see Hamilton and Karol, 1982). In our hands, a reverse phase system using a C18 column and gradient elution with acetonitrile and buffered water (pH 3.5) has yielded the best results. Reverse phase HPLC offers several advantages over straight phase HPLC. First, the separation of PGE_2 and PGD_2 is very difficult, if not impossible, with a straight phase system, but these two PGs separate by several minutes on a reverse phase system. Second, in reverse phase the unconverted arachidonic acid, which can represent a substantial amount of radioactivity, elutes after the eicosanoids, thus avoiding possible contamination of the eicosanoid peaks by tailing. Third, we have found that radiolabeled diacylglycerols can cause a significant methodological problem because they coelute on straight phase HPLC with the monohydroxy derivatives of polyunsaturated fatty acids. Apparently the major factor in straight phase columns is the interaction of the free hydroxyl group with the silica, whereas on reverse phase, the major interaction is the number of carbons. Therefore, a diacylglycerol with two fatty acids is completely separated from the hydroxylated derivative of a single fatty acid.

We have used a modification of the method of Eling and coworkers (1982) to separate radiolabeled prostaglandins and hydroxy fatty acids (Fig. 6). The radioactivity was monitored with FLO-ONE/Beta flow scintillation detector (Radiomatic) using Ready/Solv EP scintillation cocktail (Beckman) at a ratio of 2:1 (v/v), cocktail to HPLC solvent. Counting efficiency was 30% for tritium and 70% for carbon 14. Samples were dissolved in 200 μl initial conditions solvent (26% acetonitrile, 74% buffered water). Buffered water was prepared by adding 1 ml glacial acetic acid to 1 L of water and adjusting the pH to 3.5 with ammonium hydroxide (8 drops of 5 M solution). Initial conditions were maintained for 24 minutes for the elution of prostaglandins. The acetonitrile concentration was increased linearly to 40% over 3 minutes and maintained at this level for 28 minutes for the elution of dihydroxy derivatives (e.g., LTB_4). Acetonitrile was then increased to 50% over 4 minutes and maintained for 11 minutes for elution of monohydroxy derivatives. The acetonitrile concentration was then increased to 80% over 4 minutes and maintained for 10 minutes to elute free fatty acids, monoacylglycerols, diacylglycerols and triacylglycerols. Increasing acetonitrile to 100% for 10 minutes eluted the remaining glycerolipids. The column was reequilibrated at initial conditions for 15 minutes before injection of the next sample. This system provides an exel-

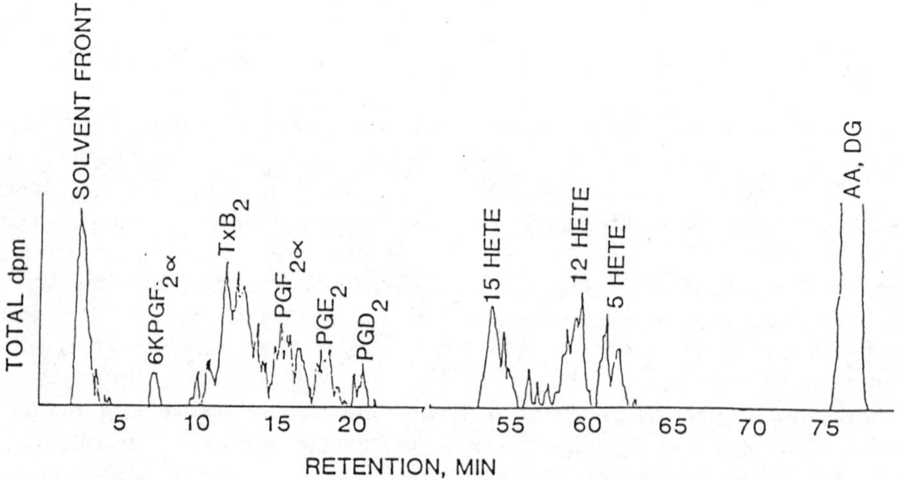

Figure 6. Separation of radiolabeled eicosanoid standards by reverse phase HPLC.

lent separation of eicosanoids. However, since the neutral lipids elute as a single peak, an alternative method, such as TLC, is needed to assess the quantity of free arachidonic acid. As mentioned above in connection with the analysis of fatty acids, preparation of the sample for injection is very important, particularly in light of the concentration of glycerolipids in the sample. For these analyses, samples are dissolved in a solvent that contains 74% water; large amounts of glycerolipid are insoluble in this solvent. This is a major problem in the analysis of eicosanoids in a fatty tissue such as brain, where the levels of glycerolipids far exceed the levels of eicosanoids. Not only do glycerolipids present a solubility problem, but extremely nonpolar glycerolipids (mainly triacylglycerols) can be difficult to elute from a reverse phase column, causing carry-over into subsequent analyses. This phenomenon can be ameliorated by using guard columns which bind the nonpolar material. An additional problem with lipid samples is the presence of protein or proteolipid that is insoluble in the injection solvent. This can be removed by filtering the sample through a 0.4 μ filter fitted onto a syringe. Also, we have used a simple precolumn extraction of the eicosanoids to remove glycerolipids (Claeys et al., 1985). Small columns of 0.5 mg silica gel (CC-40, Mallinkrodt) are prepared in Pasteur pipettes. The Pasteur pipettes are filled to 1 ml bed volume with a slurry of silica gel in diethyl ether/ acetic acid (100:0.5, by vol; solvent system C) and washed with 6 ml of the same system. After preconditioning the columns with 6 ml of n-hexane/diethyl ether/acetic acid (90:10:0.5, by vol; system A), the sample is applied onto the columns in 150 μl of system C while the column is temporarily plugged off (the solubility of the sample in system A is incomplete). After evaporation of the diethyl ether on the top of the column with the stream of nitrogen, elution is started with 6 ml of system A to elute triacylglycerols, phospholipids, cholesterol esters and part of the free fatty acids (fraction 1). A fraction containing monohydroxy acids is obtained by subsequent elution with 6 ml of a mixture of n-hexane/diethyl ether/acetic acid (60:40:0.5, by vol; system B fraction 2). Following elution of the monohydroxy acids, elution of dihydroxy acids (fraction 3) is performed with 6 ml of system C. Subsequently, elution of prostaglandins (fraction 4) is done with 6 ml of a mixture of diethyl ether/methanol/acetic acid (90:10:0.5, by vol; system D). Fractions 2, 3 and 4 are combined for HPLC analysis.

For the analysis of eicosanoids, we have found that straight phase HPLC is useful for further purification of compounds after reverse phase HPLC. Peaks from the reverse phase system are collected, converted to methyl esters, and rechromatographed on straight phase HPLC prior to further derivatization for GC-MS (Careaga and Sprecher, 1984). Chloroform:methanol is a common HPLC solvent for the separation of prostaglandins by straight phase HPLC (Hamilton and Karol, 1982). However, if flow scintillation detection is used, chloroform is an unacceptable solvent, because of quenching. Mixtures of hexane and isopropanol in various proportions (e.g., 5% isopropanol) provide a good separation of hydroxy-fatty acids, and pose no problem in terms of radioactivity detection (Birkle and Bazan, 1984a, 1984b; Careaga and Sprecher, 1984).

8. TROUBLESHOOTING

There are several important factors to consider in the collection of meaningful, accurate radioactivity data by flow scintillation detection. When confronted with relatively low levels of radioactivity, the background determination and counting efficiency of the system become critical. We have found that in addition to determination of the static background of the detector, which ranges from 30-50 cpm, determination of the flowing background is useful. This is accomplished by assessing the cpm with both scintillation fluid and HPLC solvent flowing through the detector, at the ratio and flow rate used during sample analysis. This value usually ranges from 50-70 cpm. We also use the technique of blank runs (injection of solvent blanks) to determine possible bleeding of radioactivity from the HPLC column. Depending on the age and condition of the column, this value can range from 0 to 150 cpm. We program the system to make these blank runs routinely after every four or five samples to assess carry-over of radioactivity from sample to sample and variation in the background counts.

The HPLC systems we have described for separation of fatty acids and eicosanoids contain fairly high concentrations of water, which can present a problem in terms of counting efficiency. It is necessary to use a scintillation cocktail that can handle 10-30% water, and to assess the effects of varying the water:acetonitrile ratio on counting efficiency. Using the systems we have described, there is a decrease in counting efficiency of about 2% over the course of the gradient (Fig. 7).

In the description of the various methods, we have mentioned some of the problems associated with the HPLC anaylsis of fatty acids and eicosanoids. Of primary importance in the successful application of these techniques are

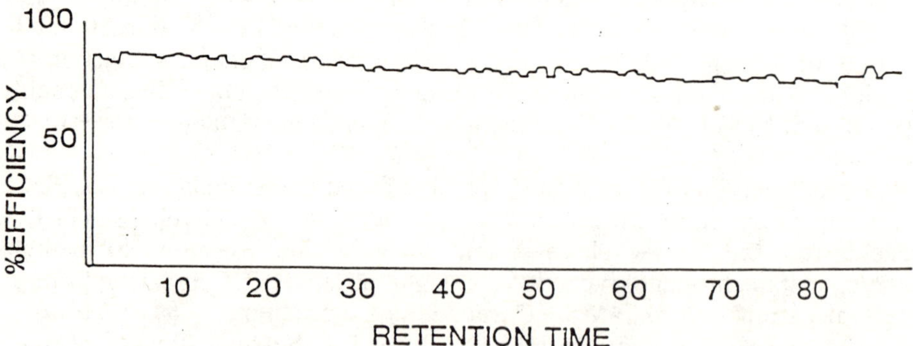

Figure 7. Counting efficiency for ^{14}C (Radiomatic FLO-ONE/Beta) obtained with the solvent system (acetonitrile:water) used for eicosanoid separation by reverse phase HPLC (Fig. 6). The discriminator setting was 1-90 (wide open window) and the subtracted background was 70 cpm.

the basic principles of all good chromatography: check the system thoroughly with standards, in a manner consistent with the sample preparation; spike actual samples with standards and verify the separation and recovery if possible, since lipids present in the actual samples can alter these parameters; characterize the system routinely, so that unusual results and deterioration of the column will be recognized quickly.

ACKNOWLEDGEMENTS

The authors' research efforts are supported by the Esther and Joseph Klingenstein Fund, Inc, New York, and the National Institutes of Health, grants EY05121, NS23002, EY04298, and EY07073.

REFERENCES

Aveldano, M.I., Horrocks, L.A. (1983). Quantitative release of fatty acids from lipids by a simple hydrolysis procedure. *J. Lipid. Res.* 24:1101-1105.

Aveldano, M.I., Van Rollins, M., Horrocks, L.A. (1983). Separation and quantitation of free fatty acids and fatty acid methyl esters by reverse phase high pressure liquid chromatography. *J. Lipid. Res.* 24:83-93.

Bazan, H.E.P., Birkle, D.L., Beuerman, R., Bazan, N.G. (1985). Cryogenic lesion alters the metabolism of arachidonic acid in rabbit cornea layers. *Invest. Ophthalmol. Vis. Sci.* 26:474-480.

Bazan, N.G. (1970). Effects of ischemia and electroconvulsive shock on free fatty acid pool in the brain. *Biochim. Biophys. Acta* 218:1-10.

Bazan, N.G., Reddy, T.S., Redmond, T.M., Wiggert, B,. Chader, G.J. (1985). Endogenous fatty acids are covalently and noncovalently bound to interphotoreceptor retinoid-binding protein in the monkey retina. *J. Biol. Chem.* 260:13677-13680.

Birkle, D.L., Bazan, N.G.(1984a). Effect of K^+ depolarization on the synthesis of prostaglandins, hydroxyeicosatetra (5,8,11,14) enoic acid (HETE) and other eicosanoids in rat retina: Evidence for esterification of 12-HETE in lipids. *Biochim. Biophys. Acta* 795:564-573.

Birkle, D.L., Bazan, N.G. (1984b). Lipoxygenase- and cyclooxygenase-reaction products and incorporation into glycerolipids of radiolabeled arachidonic acid in the bovine retina. *Prostaglandins* 27:203-206.

Birkle, D.L., Bazan, N.G. (1985). Metabolism of arachidonic acid in the central nervous system. The enzymatic cyclooxygenation and lipoxygenation of arachidonic acid in the mammalian retina. In: *Phospholipids in the Nervous System*, Vol. 2: *Physiological Roles*, L. Horrocks, J. Kanfer, G. Porcellatti (eds). Raven Press, New York, pp. 193-208.

Birkle, D.L., Bazan, N.G. (1986). The arachidonic acid cascade and phospholipid and docosahexaenoic acid metabolism in the retina. In: *Progress in Retinal Research*, Vol. 5, N. Osborne and J. Chader (eds). Pergamon Press, London, pp. 309-335.

Bleasdale, J., Hauser, G., Eichberg, J. (1985). Inositol and phosphoinositides. Proceedings of the Chilton Conference, Humana Press, New Jersey.

Careaga, M.M., Sprecher, H. (1984). Synthesis of two hydroxy fatty acids from 7,10,13,16,19-docosahexaenoic acid by human platelets. *J. Biol. Chem.* 259:14413-14417.

Claeys, M., Kivits, G.A.A., Christ-Hazelhof, E., Nugteren, D.H. (1985). Metabolic profile of linoleic acid in porcine leukocytes through the lipoxygenase pathway. *Biochim. Biophys. Acta* 873:35-51.

Corey, E.J., Niwa, H., Folch, J.R., Mioskowski, C., Arai, Y., Marfat, A. (1980). Recent studies on the chemical synthesis of eicosanoids. *Adv. Prostaglandin Thromboxane Res.* 6:19-15.

Creutz, C.E. (1981). Cis-unsaturated fatty acids induce the fusion of chromaffin granules aggregated by synexin. *J. Cell. Biol.* 91: 247-256.

Cullis, P.R., deKruijff, B., Hope, M.J., Nayar, R., Schmid, S.L. (1980). Phospholipids and membrane transport. *Can. J. Biochem.* 58:1091-1100.

Eling, T., Tainer, B., Alley, A., Warwick, R. (1982). Separation of arachidonic acid metabolites by high pressure liquid chromatography. *Meth. Enzymol.* 86:511-517.

Folch, J., Lees, M., Sloane-Stanley, G.E. (1957). A simple method for the isolation and purification of total lipids from animal tissues. *J. Biol. Chem.* 226:497-509.

Goodwin, J.S., Ceuppens, J. (1983). Regulation of the immune response by prostaglandins. *J. Clin. Immunol.* 3:295-315.

Green, K., Granstrom, E., Samuelsson, B., Axen, U. (1973). Methods for quantitative analyses of $PGF_2\alpha$, PGE_2, 9α,11s-dihydroxy-15 ketoprost-5-enoic acid, and $9\alpha,11\alpha$, 15-trihydroxyprost-5-enoic acid from body fluids using deuterated carriers and GC-MS. *Anal. Biochem.* 54: 434-453.

Hamilton, J.G., Karol, R.J. (1982). High performance liquid chromatography (HPLC) of arachidonic acid metabolites. *Prog. Lipid. Res.* 21:155-170.

Hammarstrom, S. (1983). Leukotrienes. *Ann. Rev. Biochem.* 52:355-377.

Horrocks, L.A. (1985). Metabolism and function of fatty acids in brain. In: *Phospholipids in Nervous Tissues*, J. Eichberg (ed.), pp. 173-199, John Wiley & Sons, New York.

Horrocks, L.A., Freysz, L., Toffano, G. (1986) In: *Phospholipid Research and the Nervous System: Biochemical and Molecular Pharmacology,* Springer-Verlag, New York.

Jacschik, B.A., Lee, L.H., Shuffer, G., Packer, C.W. (1978). Arachidonic acid metabolism in rat basophilic leukemia (RBL-1) cells. *Prostaglandins* 16:733-748.

Lands, W.E.M. (1979). The biosynthesis and metabolism of prostaglandins. *Ann. Rev. Physiol.* 41:633-652.

Morrison, W., Smith, L.M. (1964). Preparation of fatty acid methyl esters and di-methylacetals from lipids with boron fluoride-methanol. *J. Lipid. Res.* 5:600-608.

Pediconi, M.F., Rodriguez de Turco, E.B., Bazan, N.G. (1982). Diffusion of intracerebrally injected (1-^{14}C) arachidonic acid and (2-^{3}H) glycerol in the mouse brain. Effects of ischemia and electroconvulsive shock. *Neurochem. Res.* 7:1435-1456.

Pediconi, M.F., Rodriguez de Turco, E.B., Bazan, N.G. (1983). Effects of post decapitation ischemia on the metabolism of (^{14}C) arachidonic acid and (^{14}C) palmitic acid in the mouse brain. *Neurochem. Res.* 8:835-845.

Reddy, T.S., Bazan, N.G. (1983). Kinetic properties of arachidonoyl coenzyme A synthetase in rat brain microsomes. *Arch. Biochem. Biophys.* 226:125-133.

Rhoads, D.E., Ockner, R.K., Peterson, N.A., Raghupathy, E. (1983). Modulation of membrane transport by free fatty acids: Inhibition of synaptosomal sodium-dependent amino acid uptake. *Biochemistry* 22:1965-1970.

Rouser, G., Fleischer, S., Yamamoto, A. (1970). Two dimensional thin layer chromatographic separation of polar lipids and determination of phospholipids of phosphorus analysis of spots. *Lipids,* 5:494-496.

Saunders, R., Horrocks, L.A. (1984). Simultaneous extraction and preparation for high-performance liquid chromatography of prostaglandins and phospholipids. *Anal. Biochem.* 143:71-75.

Van Golde, L.M.G., Van den Bergh, S.G. (1977). General pathways of the metabolism of lipids in mammalian tissues. In: *Lipid Metabolism in Mammals,* F. Snyder, (ed). Plenum Press, New York. pp. 1-33.

Wolfe, L.S. (1982). Eicosanoids: Prostaglandins, thromboxanes, leukotrienes and other derivatives of carbon-20 unsaturated fatty acids. *J. Neurochem.* 38:1-13.

Progress in HPLC, Vol. 3, pp. 27-55
Parvez et al. (Eds)

HPLC methodology with flow detection of radiolabeled neutral and phospholipids, polyphosphoinositides, phosphoinositol isomers and arachidonic acid metabolites from pituitary gonadotrophs

REGINALD O. MORGAN and KEVIN J. CATT
Endocrinology and Reproduction Research Branch, NICHHD,
National Institutes of Health, Bethesda, Maryland, USA

ABSTRACT/SUMMARY

The burgeoning significance of phospholipase-catalyzed hydrolysis products of polyphosphoinositide metabolism in cell regulation has prompted the development of improved methods for their identification and quantitation. By modification of existing methods, we have utilized high resolution HPLC techniques combined with radioactivity flow detection for analysis of the principal second messengers arising from GnRH-stimulated metabolism of phosphatidylinositol-4,5-bisphosphate in rat pituitary gonadotrophs.

Single-run class separation of all major neutral lipids, phospholipids and lysophospholipids was achieved by normal-phase high-performance liquid chromatography (NP-HPLC) gradient elution with a hexane/isopropanol/water/sulfuric acid solvent system. Quantitative analysis of polyphosphoinositides required prior deacylation followed by strong anion exchange (SAX-HPLC) of the water-soluble glycerophosphoinositols using a gradient of acidic aqueous ammonium phosphate. All known inositol polyphosphates (IPs), including isomers of IP_1, IP_2, and IP_3, were resolved on a SAX-HPLC column using a more gradual gradient of ammonium phosphate. Inositol cyclic and isomeric monophosphates were even more effectively separated by weak anion exchange (WAX-HPLC) using an amine column with elution by either ammonium acetate or phosphate. Arachidonic acid and its cyclooxygenase, lipoxygenase and monooxygenase metabolites were resolved in two different systems by reverse-phase HPLC on C-18 columns with aqueous acetonitrile.

Radioactivity in HPLC effluents was quantitated by liquid scintillation counting in an on-line radioactivity flow detector (FLO-ONE/Beta Model IC, Radiomatic Instruments) which recorded raw digital data and graphed or tabulated peak integration data for one absorbance and two radioactivity channels. The automated HPLC analytical methods described here for extracts of cultured cells should have wide applicability to metabolic studies or chemical purification.

INTRODUCTION

The metabolism of inositol-containing phospholipids is receiving increased scrutiny as its involvement with intracellular signal transduction becomes apparent. Biophysical and biochemical changes within the lipid bilayer of plasma membranes are believed to contribute to cellular recognition of ligand-receptor binding and to the generation of several newly defined "intracellular messengers." These products of lipid metabolism appear to trigger a cascade of biochemical processes, including mobilization of calcium and arachidonic acid (AA) and activation of protein kinases, which culminate in specialized tissue responses such as contraction or secretion of hormones, neurotransmitters or enzymes. The orderly sequence of events and their physiological implications are discussed and illustrated, to the extent that they are currently understood, in recent review articles (Berridge, 1985; Kirk *et al.*, 1984). Novel methods for the identification and radiochemical analysis of the principal "second messengers" derived from phosphoinositides will be the focus of the present study.

The phosphoinositides are subject to initial metabolism by two major enzymatic pathways that are present in many cell types in which a rise in intracellular calcium level is a significant component of the mechanism by which those cells respond to external stimuli (Fig. 1). One of these pathways features a PHOSPHOLIPASE C-catalyzed hydrolysis of phosphatidylinositol-4,5-bisphosphate (PIP_2) at the "sn-3" glycerol carbon to yield diglyceride (DG) plus inositol-1,4,5-trisphosphate (IP_3). The other is a PHOSPHOLIPASE A2-catalyzed hydrolysis of unsaturated fatty acid-rich phosphatides of inositol, ethanolamine and choline (PI, PE and PC) at the sn-2 glycerol carbon to yield a 2-lysophospholipid (LPL) plus a fatty acid, often AA. It should be recognized that these two pathways are not mutually exlusive; rather they are functionally intertwined, as the products of one pathway may regulate or participate in the other.

The immediate products of lipid hydrolysis have been shown to exert prominent effects on calcium mobilization (viz. IP_3 and AA), protein kinase C activation (eg., DG) and plasma membrane properties (LPL and AA). However, a knowledge of their further metabolism is required to understand the dynamics of their regulation and to identify additional products, which may also regulate cell function. Thus, areas to be explored include the identification of dephosphorylation (or phosphorylation?) products of IP_3 metabolism and the enzymatic oxidation products of AA metabolism, as well as the fate of DG metabolism via DG kinase to phosphatidic acid or via DG lipase to monoglyceride and, subsequently, free arachidonate. Identification and quantitation of each of the components within these distinct classes of compounds require specific techniques for cell-labeling, extraction, sample processing, separation (see Fig. 1) and detection.

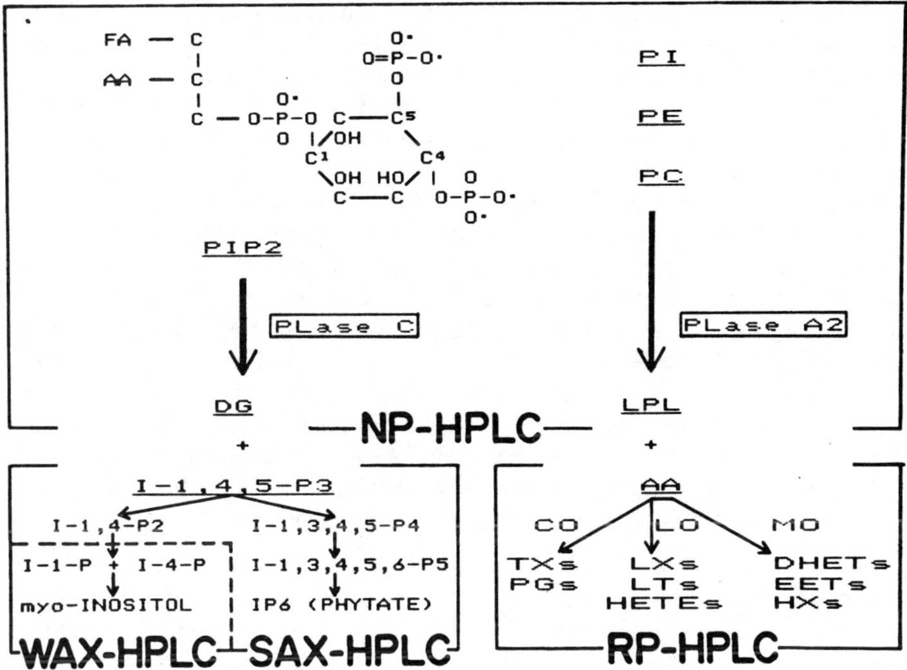

Figure 1. Activation of phospholipases C and A_2 as mechanisms for GnRH-stimulated phospholipid hydrolysis leading to the production of inositol polyphosphates, diglyceride, lysophospholipids and arachidonic acid metabolites. Modes of product analysis include NP-HPLC of neutral lipids, phospholipids and lysophospholipids; SAX-HPLC of all inositol polyphosphates; WAX-HPLC (on aminopropyl-bonded silica) for improved resolution of inositol monophosphates, and RP-HPLC (on C18-bonded silica) of arachidonic acid and its metabolites.

Notwithstanding the relatively recent appreciation of the importance of polyphosphoinositide metabolism in cell function, much of our biochemical knowledge about the key intermediates (Fig. 1) derives from detailed analytical studies over the past several decades (eg., Brockerhoff and Ballou, 1961; Ellis *et al.*, 1963; Grado and Ballou, 1961; Hendrickson and Ballou, 1964; Tomlinson and Ballou, 1961). The techniques employed have evolved with the aims of achieving more complete and definitive resolution, greater efficiency of time and materials, and enhanced sensitivity and precision for quantitative analyses. For example, inositol polyphosphates have been resolved by paper chromatography over 7-10 days (Grado and Ballou, 1961) by high voltage ionophoresis in electrophoresis tanks (Frahn and Mills, 1959), by gradient elution from large volume anion exchange columns (Ellis *et al.*, 1963), by step-wise elution from ion exchange resin minicolumns (Berridge *et al.*, 1983), and by gradient elution with "fast" performance liquid chromatography (Irvine *et al.*, 1984). These methodological developments have introduced progressively more sophisticated approaches to the analysis of phospholipid metabolites, and provide the basis for the further innovations described below.

The literature is replete with thin-layer chromatographic methods for phospholipid analysis but the recent application of HPLC has met with only limited success (Aitzetmuller, 1982; Jungalwala *et al.*, 1984; Kaduce *et al.*, 1983; Patton *et al.*, 1982; Yandrasitz *et al.*, 1981). On the other hand, HPLC has set the standard for analysis of arachidonic acid metabolites during a period of intense study this past decade, but most individual studies have limited their scope to select classes of compounds (Osborne *et al.*, 1983; review by Hamilton and Karol, 1982). These include cyclooxygenase products such as prostaglandins (PGs) and thromboxanes (TXs), lipoxygenase products such as leukotrienes (LTs), hydroxyeicosatetraenoic acid (HETEs) and lipoxins (LXs), or monooxygenase products such as epoxyeicosatrienoic acids (EETs) and hepoxilins (HXs). One of the chief aims in developing HPLC methods for lipid analysis has therefore been to achieve adequate separation of as many potential intermediates as possible so as not to confound the accurate measurement of those that may be of particular interest.

Four of the most significant recent developments which have facilitated studies of cellular lipid metabolism are pertinent here:

1. RADIOISOTOPES for virtually all of the structural components of phospholipids have become available for the study and quantitative analysis of specific metabolic pathways. This obviates the special problems of analyzing diradylglycerides and phosphoinositols, which lack suitable spectrophotometric properties for convenient in-line detection, and would normally require chemical analysis subsequent to separation.

2. CELL CULTURE TECHNOLOGY for the maintenance of primary cultures of rat pituitary cells in the presence or absence of serum has enabled the incorporation of radiolabeled metabolic precursors to full equilibrium conditions. This has not been generally possible with most cell types used to study phosphoinositide metabolism but it has been an important advantage in achieving a quantitative assessment of target cell responses, which closely reflect actual mass changes in lipid intermediary metabolites.

3. HIGH-PERFORMANCE LIQUID CHROMATOGRAPHY (HPLC) systems now provide efficient, high resolution separations of various classes of compounds on a wide range of column packings. They also offer the capacity for full automation of sample processing and in-line, nondestructive analyses. While adequate solvent separation systems have only recently been developed for the resolution of complex lipids, normal and reverse-phase HPLC methodology has always provided the most effective means of analyzing the myriad of arachidonic acid metabolites.

4. RADIOACTIVITY FLOW DETECTORS with computer aided data acquisition and reprocessing capabilities offer a rapid and effective method for quantitative radiochemical detection of compounds of interest. This has been of special value in the present studies, where HPLC separation techniques were being developed and where the alternative of chemical analysis of individual HPLC fractions would have been a substantial hindrance.

SAMPLE PROCESSING TECHNIQUES

Cell culture and radiolabeling

Suspensions of rat anterior pituitary cells were prepared from excised adenohypopheses of 50 day old female rats using a trypsin-dispersion technique (Hyde *et al.*, 1982). The suspension generally showed 80%-90% viability, with gonadotrophs comprising approximately 10%-12% of the cell population. Cell suspensions enriched up to 60% in gonadotrophs were obtained by centrifugal elutriation of the pituitary cell suspension (Hyde *et al.*, 1982). Cells were cultured with Cytodex-1 beads (Pharmacia) at a ratio which would provide anchorage for about 500 cells/bead in a 6 cm plastic Petri dish. The culture medium for radioisotope incorporation was Medium 199 (M199) with Earle's salts containing 25 mM Na-Hepes buffer (pH 7.4), 100 U/ml penicillin G, 100 μg/ml streptomycin and, after the first day of incubation, 10% horse serum. During radiolabeling with inositol, arachidonate or phosphate, media were employed which lacked, respectively, inositol, phenol red or phosphate. Cell-bead suspensions were generally incubated for one day at 10 million cells per ml of serum-free M199 containing 1-10 μCi/ml radiolabeled lipid precursors at 37°C under 5% CO_2 in air in a humidified tissue culture incubator (Forma Scientific). The incubation was continued for a further one to two days with an additional volume of medium and 10% horse serum to incorporate radiolabels approaching equilibrium conditions for [^{3}H]- or [^{14}C]-glycerol, [^{3}H]- or [^{14}C]-myo-inositol, [^{3}H]- or [^{14}C]-arachidonic acid and [^{33}P]- or [^{32}P]-phosphate (New England Nuclear).

Extractions and deacylation

Following pharmacologic treatment of prelabeled cells, reactions were stopped by addition of ice-cold acids and/or organic solvents, depending on the subsequent analysis. For phosphoinositol and phospholipid analysis, trichloroacetic acid was added to 5% final concentration. After sonication and centrifugation, the supernatant was removed and combined with two subsequent water washes of the cell pellet. This aqueous phosphoinositol fraction was extracted with 4 x 2 ml diethyl ether and the remaining acid was neutralized with ammonium hydroxide. Phospholipids in the cell pellet were extracted with chloroform/methanol/HCl as described by Creba *et al.* (1983).

For "cyclic" inositol phosphate and phospholipid analyses chloroform, methanol and 1 mM EDTA in water were added for a final volume ratio of 2.5/1.25/0.75 to produce a two-phase system. The upper phase was removed and washed once with 1.5 ml chloroform/methanol = 85/15; the lower phase was washed once with methanol/water = 1/1. Both upper phases were combined and extracted again with chloroform/methanol = 85/15 to minimize the loss of triphosphoinositide. All organic phases were combined, dried under a nitrogen stream and re-extracted for phospholipids under acidic conditions as above (Creba *et al.*, 1983).

For arachidonic acid metabolites in aqueous culture medium, formic acid was added to pH 3 and lipids were extracted onto a methanol- and water-washed C-18 SEP-PAK (Waters) followed by elution with methanol. Alternatively, for simultaneous extraction of phospholipid-bound and nonesterified AA together, 3 ml ice-cold hexane/2-propanol = 2/1 was mixed and sonicated with 0.2 ml cell suspension (Saunders and Horrocks, 1984). Filtered organic extracts of intact phospholipids were dried under a nitrogen stream and redissolved in 200 μl of hexane/2-propanol/water = 98/1.9/0.1 for HPLC injection. For analysis of inositol or phosphate-labeled phosphoinositides, dried lipid samples were deacylated according to the method of Clarke and Dawson (1981) and the resulting water-soluble glycerophosphodiesters were analyzed by strong anion exchange HPLC as described below.

Precursors and chromatographic standards

Nonradioactive lipid standards for ultraviolet detection were obtained from Sigma (viz. prostaglandins D_2, E_2 and $F_2\alpha$, phosphatidylglycerol, diphosphatidylglycerol, phosphatidylserine, phosphatidylethanolamine, phosphatidylinositol, phosphatidylcholine and sphingomyelin). Dioleoylglyceride and 2-arachidonoyl phosphatidic acid were obtained from Avanti Polar Lipids, arachidonic acid was from NuChek Prep and Merck Frosst Canada provided leukotrienes B_4, C_4, D_4, and E_4, 15-HETE, 12-HETE and 5-HETE.

Authentic, radiolabeled precursors from New England Nuclear included [^{3}H]- and [^{14}C]-glycerol, [^{3}H]- and [^{14}C]-myo-inositol, [^{3}H]- and [^{14}C]-arachidonic acid and [^{32}P]- or [^{33}P]-phosphoric acid. Other tritium-labeled standards from NEN included inositol-1-monophosphate (Ins-1-P), inositol-1,4-bisphosphate (Ins-1,4-P_2), inositol-1,4,5-trisphosphate (Ins-1,4,5-P_3), phoshatidylinositol (PI), phosphatidylinositol-4-phosphate (PIP), phosphatidylinositol-4,5-bisphosphate (PIP_2), 5-, 12- and 15-hydroxyeicosatetraenoic acids (5-, 12- and 15-HETEs).

Certain radiolabeled standards were prepared by enzymatic or chemical treatment of pure lipids or prelabeled cells (Mangold *et al.*, 1984). These included phospholipase C treatment of [^{3}H]-glycerol, arachidonate or inositol phosphoinositides to produce labeled diglyceride and inositol phosphates ([^{3}H]-IP_1, IP_2 and IP_3). Alkaline hydrolysis in monomethylamine (Clarke and Dawson, 1981) of inositol-labeled phosphoinositide standards was used to produce the corresponding glycerophosphoinositols "GPI, GPIP or $GPIP_2$" without dephosphorylation artifacts. Inositol-1:2-cyclic phosphate (Ins-1:2-cP) was produced by a 2 h incubation of [^{3}H]-PI with unlabeled PE (70 μg/ml) plus phospholipase C from *B. cereus* (1000 U/ml) in 0.1 ml buffer (pH 5.5) consisting of 2[N-morpholino]ethanesulfonic acid (MES) 50 mM, KCl 80 mM, $CaCl_2$ 1 mM and BSA 0.5 mg/ml, followed by neutral aqueous extraction (above). Ins-1-P and Ins-2-P were then formed by heating the Ins-1:2-cP in 0.1 *N* HCl at 100°C for 15 minutes (Binder *et al.*, 1985).

[^{3}H]-Inositol-4-phosphate (Ins-4-P) standard was produced by strong base hydrolysis of authentic [^{3}H]-inositol-labeled phosphatidylinositol-4-phosphate (PIP) in 2 *N* KOH for 30 minutes, followed by removal of K^+ ions on cation exchange resin (Bio-Rad AG 50W-X12, H^+ form), ether extraction of lipids and neutralization with ammonium hydroxide (Grado and Ballou, 1961). The inositol monophosphate ester product has previously been characterized as Ins-4-P, distinguishable from other monophosphate isomers (Grado and Ballou, 1961), and formed by a process not involving cyclic intermediates or phosphate group migration (Tomlinson and Ballou, 1961).

HPLC and detection systems

HPLC instrumentation was from Waters and consisted of a Model 710B "WISP" auto-injector, three Model 6000A pumps, a Model 680 automated gradient and timed event controller, one Model 441 fixed wavelength absorbance detector, one Model 270 LC variable wavelength spectrophometer and a Model M730 data module (for absorbance data graphing and tabulation). Arachidonate metabolites were initially analyzed by reverse-phase HPLC on a Novapak C18 (4μ) cartridge, 0.8 x 10 cm in a Z-module Radial Compression Module (Waters). A more recent method (described below) employed a 0.46 x 15 cm Adsorbosphere C18 3-micron column (Alltech Applied Science) with aqueous, acidic acetonitrile gradients. All other analyses described here were performed on 0.46 x 25 cm columns of Alltech Adsorbosphere with 5μ particle packing (silica, SAX or NH_2), protected before the injector by a solvent-saturating precolumn of pellicular silica and postinjector by a prefilter and universal direct connect cartridge guard column of 0.46 x 1.0 cm (Alltech) containing the same packing material as the analytical column.

Radioactivity in HPLC effluants from analytical columns was detected, after absorbance monitoring, by continuous liquid scintillation counting in FLO-SCINT cocktails with a Radiomatic FLO-ONE/Beta Model IC flow detector. The different counting mixtures consisted of FLO-SCINT I:HPLC organic eluant = 3:1.5 ml/min for NP-HPLC of phospholipids; FLO-SCINT II:aqueous acetonitrile or alcohol eluant = 4:1 ml/min for RP-HPLC of arachidonate metabolites; FLO-SCINT III:buffer eluant = 6:1 ml/min for SAX-HPLC of inositol phosphates. These were generally counted in 1.0 or 2.5 ml flow-cells over 6-second update times within appropriate discriminator settings derived from scintillation spectrum analysis using the FLO-ONE/Beta detector and software. There is one important technical note for those who may wish to increase the hexane content of the initial NP-HPLC solvent described below in order to achieve full separation of the early eluting neutral lipids: The sulfuric acid in the mobile phase may induce microprecipitation due to the scarcity of water in the effluant-cocktail mixture, but this can be fully obviated by mixing the HPLC effluant with FLO-SCINT III containing 1% water by volume (instead of the purely organic FLO-SCINT I).

Graphic and tabulated radioactivity data were calculated in disintegrations per minute (dpm) corrected for predetermined values of background, gradient-related changes in counting efficiency and channel crossover by FLO-ONE/Beta Program software (version #4e.012.11110) on a Micromate Personal Micro-Computer within the FLO-ONE/Beta detector. This was facilitated in some cases with the use of a High Level Integration software package, version #4e.012.00010 (Radiomatic), which also allowed for varying degrees of a 3-point data smoothing function in addition to the standard 5-point weighted average calculation for 1.0 and 2.5 ml flow-cells. In later runs, the UV absorbance signal from the Model 270 LC detector was corecorded with radioactivity data by the FLO-ONE/Beta Internal Computer with the connection of an analog-to-digital signal processing board.

RESULTS AND DISCUSSION

Mobile phase gradient programs

Final versions of the solvent mixtures (SM) and gradients employed are summarized below. All solvents were HPLC-grade (Fisher) including water, which was distilled, deionized and filtered free of organic and particulate matter. Mixtures were degassed by sonication or stirring under vacuum and helium-sparged during use. All gradients referred to were linear.

I. *Normal-phase (NP-) HPLC of neutral and phospholipids:*

	% Hexane	% 2-Propanol	% Water	% H_2SO_4
SM-A:	98.99	1	0	0.01
SM-B:	68.75	30	1.125	0.125
SM-C:	0	100	0	0

Column: Adsorbosphere 5μ silica, 0.46 x 25 cm steel (Alltech).
Mobile phase flow rate: 1.5 ml/min.
Sample injection solvent: Hexane/2-propanol/water = 98/1.9/0.1.

"NLPL" Gradient Program I: 100% SM-A at 1.5 ml/min isocratic for 5 minutes; linear gradient to 27% SM-A + 73% SM-B between 5-60 minutes; held isocratic from 60-90 minutes; and (optional) linear gradient to 100% SM-B between 90-110 minutes. The post-run wash and reequilibration consisted of 100% SM-C at 1 ml/min between 111-125 minutes; linear to 100% SM-A (or pure hexane) over 125-130 minutes; increase to 3 ml/min SM-A between 130-135 minutes; isocratic at 3 ml/min SM-A from 135-150 minutes; switched to 1.5 ml/min SM-A prior to next 200 μl injection, defaulting to 0 ml/min at 160 minutes in the absence of sample. For the final daily wash, 2-propanol was replaced with 5% water in 2-propanol and the column was equilibrated for storage in 100% hexane.

Gradient program I served to resolve all major neutral and phospholipids as monitored by flow scintillation counting applied to four different radiolabels (Figs. 2A, 2B, 2C, 2D). The inclusion of [^{3}H]-glycerol in the pituitary cell culture medium over a three day incubation distributed this radiolabel evenly throughout all glycerolipids (Fig. 2A), such that radiochemical quantitation accurately reflected the relative mass of individual lipid classes (Table 1). [^{3}H]-Glycerol-labeled PC (as two peaks, the latter containing sphingomyelin also) and TG were the major lipids detected, with PS, PE and PI each

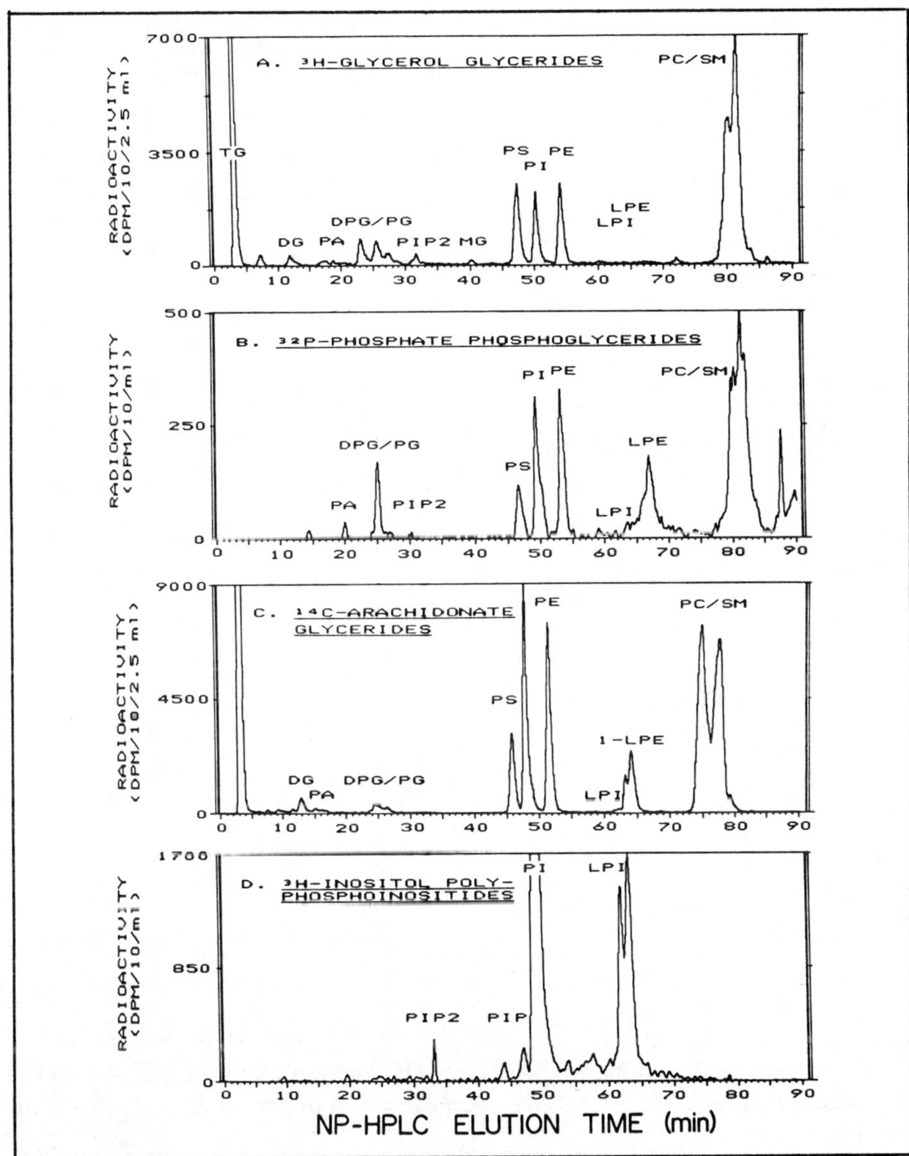

Figure 2. NP-HPLC separation by gradient program I of neutral lipids and phospholipids extracted from cultured anterior pituitary cells. Cells were cultured on Cytodex-1 beads at 500 cells per bead for 3 days in M199 containing the indicated radiolabel in order to achieve incorporation approaching equilibrium conditions. Details of culture, extraction and analysis are given in the section on Sample Processing Techniques, and a quantitative summary and list of abbreviations is given in Table 1.

A: [^{3}H]-Glycerol, 5 μCi per 0.7 $\times$ 10^6 gonadotroph-enriched anterior pituitary cells (APC), with 3.5% uptake into lipid sample; B: [^{32}P]-Phosphate, 8 μCi per 1.9 $\times$ 10^6 APC, with 0.2% uptake into lipid extract (in phosphate-containing M199); C: [^{14}C]-Arachidonate, 1 μCi per 10 $\times$ 10^6 APC, with 32% uptake into lipid sample; D: [^{3}H]-Myo-inositol, 2 μCi per 0.75 $\times$ 10^6 gonadotroph-enriched APC, with 5.4% uptake into lipid sample.

Table 1. Percentage distribution of various radiolabeled precursors in neutral and phosphoglycerides of cultured anterior pituitary cells after a 3-day equilibration period. Radiolabels included [^{3}H]-glycerol, [^{32}P]-phosphoric acid, [^{14}C]-arachidonic acid and [^{3}H]-myo-inositol. NP-HPLC profiles are displayed in Figures 2A-D.

Lipid	Abbrev.	RT (min)	[^{3}H]-G (%)	[^{32}P]-P (%)	[^{14}C]-AA (%)	[^{3}H]-I (%)
1. Cholesterol esters/ Triglycerides/TG	CE/TG	3.5	19.0	—	33.4	—
2. Fatty acid metabolites (eg. HETEs)	FAM	5–10	—	—	0.8	—
3. ?	?	8.3	0.9	0.4	—	—
4. Diradylglycerides	DG	14.0	1.2	—	2.5	—
5. Phosphatidic acid	PA	17.0	0.9	0.8	0.5	—
6. Diphosphatidylglycerol	DPG	25.1	2.9	4.0	0.9	—
7. Phosphatidylglycerol	PG	26.1	2.7	0.7	0.3	—
8. Phosphatidylinositol -4,5-bisphosphate	PIP_2	32.3	1.2	0.3	0.8	0.6
9. Monoglycerides	MG	40.1	0.5	—	0.3	—
10. Phosphatidylserine	PS	47.2	8.2	4.0	4.1	—
11. Phosphatidylinositol -4-phosphate	PIP	47.5	?	?	?	1.9
12. Phosphatidylinositol	PI	49.3	6.3	9.3	10.4	80.7
13. Phosphatidylethanolamine	PE	53.2	7.6	9.2	9.0	—
14. Lyso-phosphatidylserine	LPS	62.0	0.4	1.5	0.3	—
15. 2-Lyso-phosphatidylinositol	LPI	64.0	0.5	4.7	1.3	16.8
16. 1-Lyso-phosphatidyl-ethanolamine	LPE	65.3	0.5	7.6	3.5	—
17. Phosphatidylcholine	PC	76.2	19.2	—	16.9	—
18. Phosphatidylcholine & sphingomyelin	PC/SM	77.2	27.3	44.2	16.7	—
19. 1-Alkyl-2-acetyl-phosphatidylcholine	"PAF"					
20. 1/2-Lyso-phosphatidylcholines	LPC		0.4	13.3	?	—
TOTAL dpm			328158	34048	712941	201813

representing approximately 7% of the total activity (see Table 1 for abbreviations). Among the PI cycle intermediates, DG and PA eluted as distinct peaks representing only 2.5% and 0.9%, respectively, of the total [^{3}H]-glycerol activity.

The method was adapted from that developed originally by Yandrasitz *et al.*, (1981, 1983). Modification of the solvent mixtures and gradient programs was, however, required to resolve the neutral lipids and PA from the solvent front and to achieve baseline separation of phospholipids and lysophospholipids. Elution profiles have been easily reproducible in over 100 runs, although adequate equilibration time (ca. 10-15 minutes at 3 ml/min SM-A) was important for consistent separation of the early eluting DG and PA species. Absorbance detection at 205 nm was helpful in the identification of unlabeled lipid standards, but of limited value for quantification.

Pituitary cell phospholipids could be selectively labeled by inclusion of

[^{32}P]-phosphoric acid in the culture medium, preferably with reduced concentrations of unlabeled phosphate present (Fig. 2B). [^{33}P]-phosphoric acid was an effective substitute which, although more expensive, eliminated the external radiation hazard accompanying [^{32}P]. Both labels could be detected by flow scintillation counting with similar efficiencies of ca. 90% in a wide-open counting window, or 60% in a window which completely excluded [^{3}H] energy activity. Use of this radiolabel specifically enhanced the detection of PA, DPG and the lysophospholipids, including 1-lysoPE and, eluting after PC, lysoPC and 1-alkyl-2-acetylPC (platelet-activating factor).

Arachidonoyl-containing neutral and phospholipids were most efficiently labeled with [^{3}H] or [^{14}C]-arachidonic acid (Fig. 2C), with two peaks of PC showing the greatest incorporation. PI, PE and its plasmalogen (latter measured as 1-LPE under acidic conditions) and DG were found to be especially rich in AA-containing species, particularly after cell stimulation with GnRH. Note that neutral lipid subclasses including monohydroxy fatty acid metabolites of AA can, after methylation, be even more fully resolved in a similar starting solvent with a slightly lowered isopropanol concentration (Hamilton and Comai, 1984).

Expanding research into the dynamics of polyphosphoinositide metabolism made it of interest to follow the incorporation of [^{3}H]-myo-inositol into this important class of lipids (Fig. 2D). While PI and its lyso derivative eluted as easily detectable peaks, PIP and PIP_2 did not always elute as sharp peaks with reproducible retention times. Thus, PIP was not always resolved from PI, or PS when other radiolabels were used. PIP_2 was observed to elute, paradoxically, before other phosphoinositides and, depending on the extraction procedure, varied considerably in elution time.

The problem of peak resolution and quantitation of the polyphosphoinositides was approached by improving the lipid sample preparation procedure, as recommended by Brockerhoff (1963). Thus, conversion of PIP_2 from its divalent cationic salt to the triethylammonium salt improved its solubility characteristics and produced a larger, sharper peak at 25 minutes in gradient program I. (Fig. 3A). However, as shown in Fig. 3B, the triethlyammonium-EDTA wash was not consistently effective in suppressing the broad band of PIP_2 observed between PI and lysoPI nor in eliminating the late eluting Ca/Mg-PIP_2 peak. The procedure was, however, essential for the quantitative deacylation of phosphoinositides (Clarke and Dawson, 1981) preliminary to their analysis as glycerophosphoinositols, to be discussed next.

II. Strong anion exchange (SAX-) HPLC of inositol polyphosphates

SM-A: HPLC-grade water.
SM-B: 1 *M* HPLC ammonium phosphate, to pH 3.35
with HPLC phosphoric acid.
Column: Adsorbosphere 5μ SAX, 0.46 x 25 cm steel (Alltech).
Mobile phase flow rate: 1.0 ml/min.
Sample injection: 0.2-2 ml of a neutral, low-salt, aqueous solution.

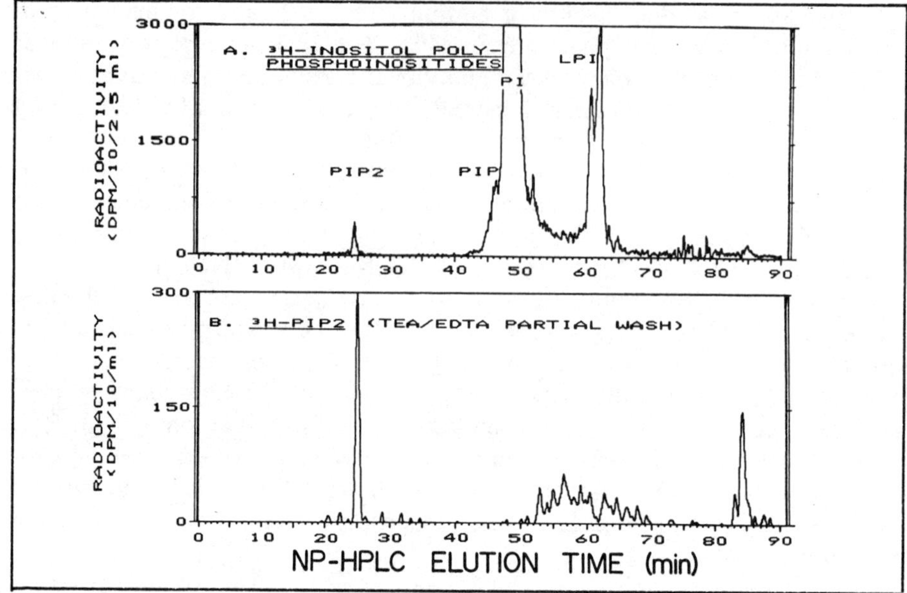

Figure 3. Illustration of problems with retention time and resolution of polyphosphoinositides during NP-HPLC analysis by gradient program I.

A: [^{3}H]-Inositol-labeled anterior pituitary cell lipid extract showing relatively early elution of a portion of PIP_2 at 25 minutes, poor separation of PIP from PI (and PS), additional peak and unresolved radiolabel between PI and lyso-PI, and additional PIP_2 peak at 85 minutes; B: [^{3}H]-Inositol- labeled PIP_2 standard (NEN) following incomplete wash of dissolved lipid with an equal volume of 0.1 M triethylammonium-EDTA (pH 7) (intended to remove divalent cations). Note the sharp peak of "salt free" PIP_2 at 25 minutes, the broad band of original sample between 52-70 minutes, and the late peak at 85 minutes representing divalent cation salt(s) of PIP_2.

"IP" Gradient Program II: The standard program for complete resolution of inositol polyphosphate isomers on an unused column began with 100% SM-A at 1 ml/min for the initial 5 minutes. This was followed by a linear gradient to 0.9 *M* ammonium phosphate (90% SM-B) at 1ml/min between 5-95 minutes, i.e., a concentration rate change of 10 mM/min. This rate was increased to 15 mM/min (still up to 0.9 *M* ammonium phosphate) when resolution was not critical, as for separation of glycerophosphoinositol products of polyphosphoinositide deacylation. The rate of salt gradient rise was decreased to 5 mM/min when column sample retention had markedly deteriorated after 100 runs. At the end of the run, effluant was restored to 100% SM-A within 5 minutes and the column re-equilibrated with 100% SM-A at 1.5 ml/min for 15 minutes before adjusting flow to 1 ml/min SM-A prior to the next automatic injection.

The deacylation procedure for [^{3}H]-inositol-labeled phosphoinositides (Clarke and Dawson, 1981) was carefully studied to ascertain that it was a quantitatively complete reaction that was free of artifactual products such as dephosphorylated derivatives. SAX-HPLC analysis of the GPI, GPIP and $GPIP_2$ derived from cellular phosphoinositide extracts washed with triethylammonium-EDTA resulted in three sharp, distinct peaks of radioactivity (Fig. 4). Despite the extra preparative step involving deacylation, this method provided the simplest, most reliable, quantitative analysis of cellular and standard phosphoinositides when phospholipase C-catalyzed metabolic changes were being studied. The normal-phase HPLC analysis (Figs. 2D and 3) was reserved for studies involving phospholipase A_2-induced changes in PI and lyso-PI levels and for analysis of major neutral and phospholipid classes.

The possible utility of different radiolabels for validating lipid analysis by the deacylation/anion exchange method was investigated in [^{3}H] plus [^{32}P] double-labeled phospholipids (Fig. 5). With correction for channel crossover, [^{3}H]-inositol GPI's could be compared with the elution position of

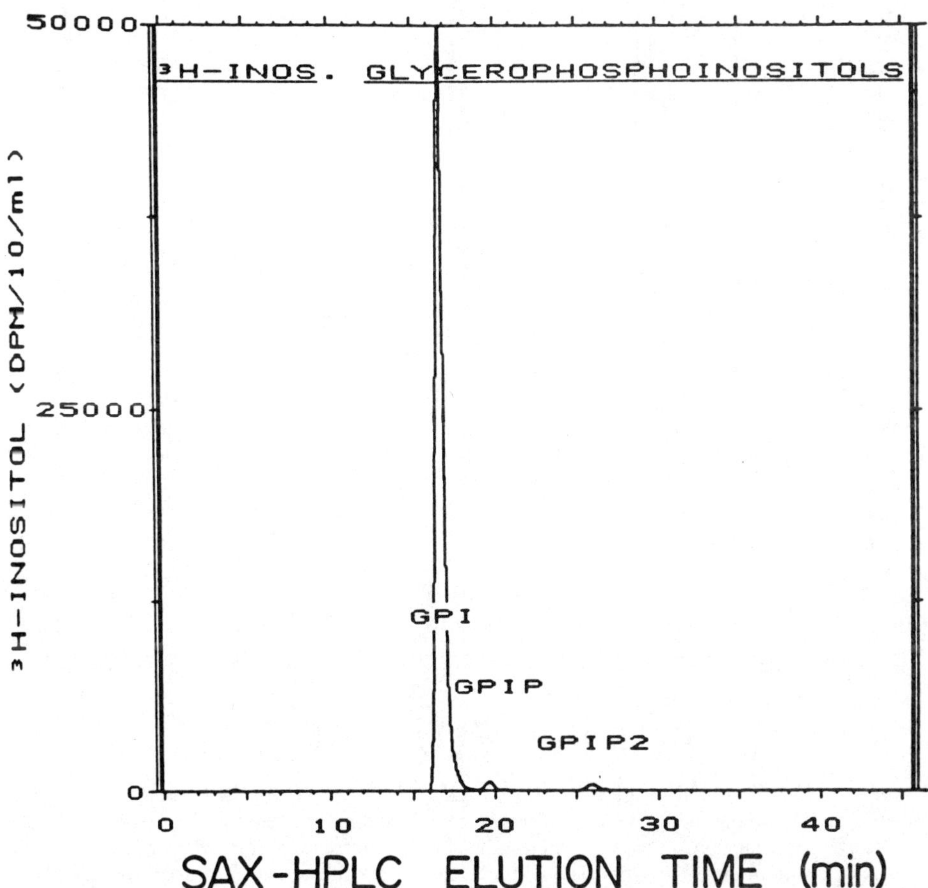

Figure 4. SAX-HPLC resolution of [^{3}H]-inositol-labeled glycerophosphoinositols derived from cellular polyphosphoinositides. [^{3}H]-Inositol-labeled polyphosphoinositides extracted from gonadotroph-enriched APC were washed with 0.1 *M* triethylammonium-EDTA (pH 7), deacylated by monomethylamine, and the neutral aqueous extract was eluted from a SAX-HPLC column with ammonium phosphate (pH 3.35) by gradient program II (viz. 0-0.6 *M* between 5-45 minutes). The distribution of the 300K sample dpm was 97.3% in GPI, 1.5% in GPIP and 1.2% in $GPIP_2$.

[^{32}P]-glycerophosphate esters derived from total phosphoglycerides. Calculation of the [^{32}P]/[^{3}H] ratios established that GPI and $GPIP_2$ could be resolved from other lipid esters, but that GPIP coeluted with other, unidentified [^{32}P]-labeled glycerophosphate esters. Double-labeling with [^{3}H] plus [^{14}C]-glycerol also produced glycerophosphate esters in both counting channels (Fig.6) and confirmed that GPI and $GPIP_2$, but not GPIP, were well-resolved for accurate quantitation using any of three radiolabels tested.

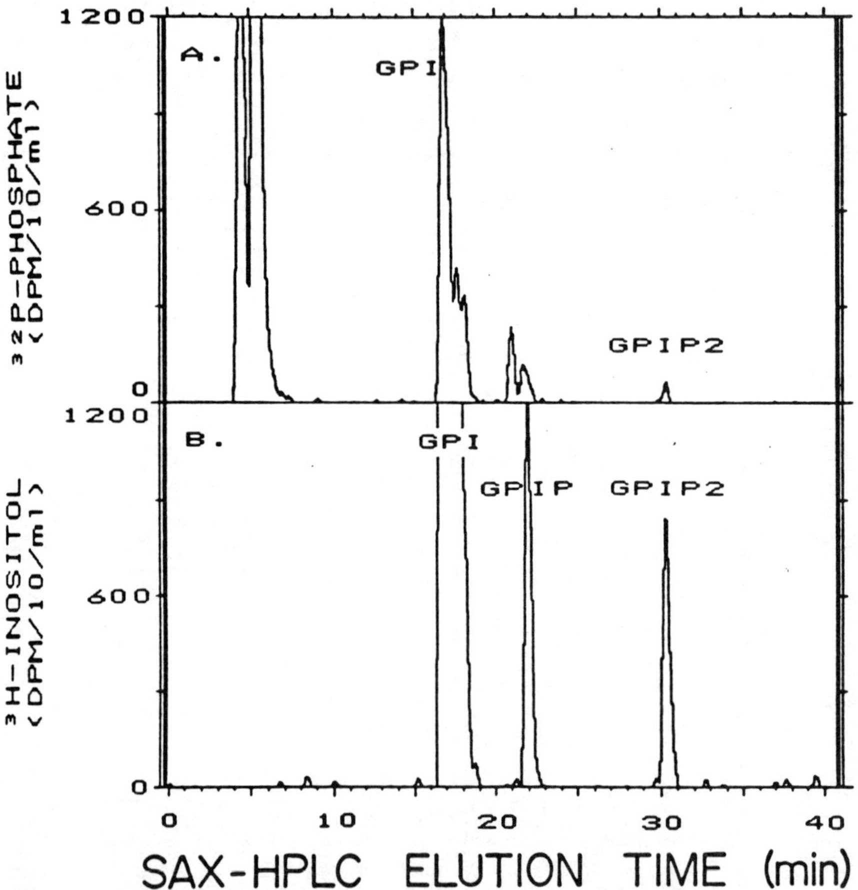

Figure 5. SAX-HPLC resolution of [^{3}H]-inositol plus [^{32}P]-labeled phosphoglycerides deacylated by monomethylamine treatment of a phospholipid extract from gonadotroph-enriched APC labeled three days to equilibrium. Elution was by gradient program II with 0-0.6 *M* ammonium phosphate (pH 3.35) between 5-45 minutes. Panel A shows the distribution of 50K dpm into [^{32}P]- glycerophosphate esters of choline and ethanolamine around 5 minutes, with glycerophosphoglycerol, glycerophosphate and glycerophosphoserine eluting between GPI (at full-scale) and GPIP. Panel B shows the distribution of 500K dpm in the [^{3}H]-inositol-labeled "glycerophosphoinositols" GPI, GPIP (at full-scale) and $GPIP_2$. The ratios of [^{32}P]/[^{3}H] in the GPIs indicate that the large [^{32}P] peak at 17 minutes may be entirely GPI, that $GPIP_2$ is separate from other lipids, but that the [^{32}P]-GPIP peak contains some additional glycerophosphate ester.

SAX-HPLC analysis of inositol phosphate isomers proved to be the most successful application of the method to the study of phosphoinositide metabolism. The use of an extended gradient for the analysis of water-soluble [^{3}H]-phosphoinositide metabolites led to the unexpected finding of inositol phosphates eluting after IP_1, IP_2 and IP_3 in either mixed pituitary cells (Fig. 7A) or elutriated gonadotrophs (Fig. 7B). Such species have previously been identified only in nonmammalian red blood cells (Bartlett, 1982; Isaacks *et al.*, 1982).

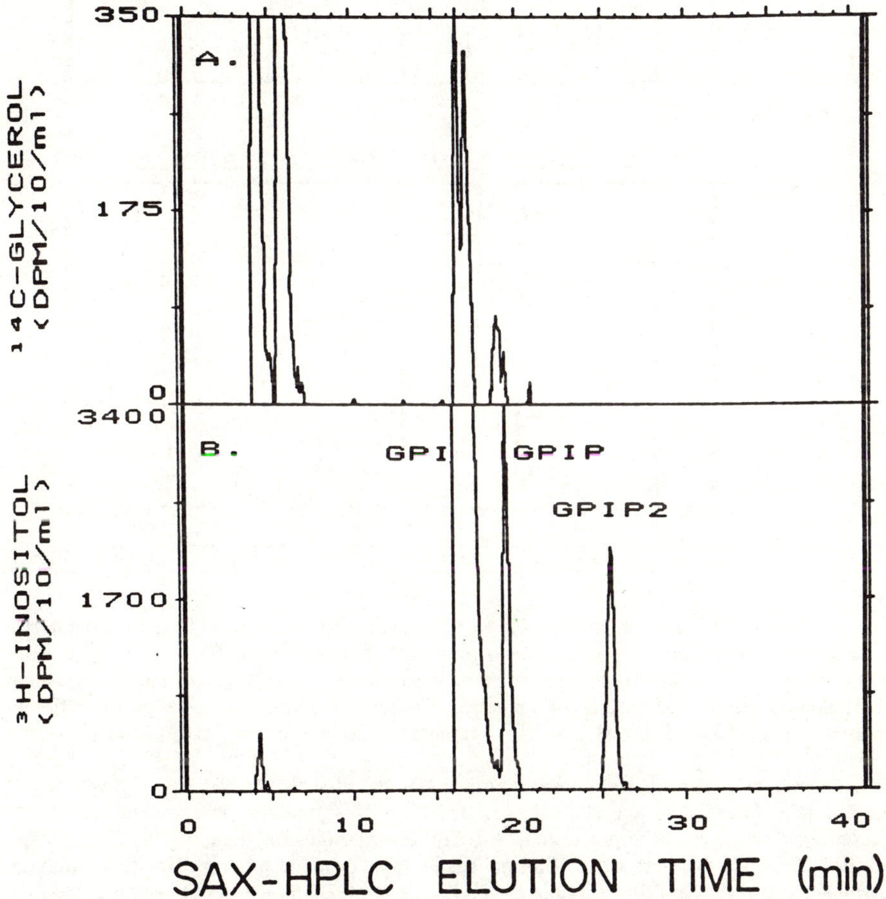

Figure 6. SAX-HPLC resolution of [^{3}H]-inositol plus [^{14}C]-glycerol double-labeled glycerophosphate esters obtained by deacylation of phospholipids from anterior pituitary cells cultured three days in the presence of both radiolabels. Elution was by gradient program II with 0-0.6 *M* ammonium phosphate (pH 3.35) between 5-45 minutes. Panel A shows the distribution of 17K dpm [^{14}C]-glycerol into glycerophosphocholine and glycerophosphoethanolamine (at 5 minutes) with glycerophosphoglycerol, glycerophosphate and glycerophosphoserine eluting between GPI (at full-scale) and GPIP. Panel B shows the distribution of 1000K dpm [^{3}H]-inositol in the glycerophosphoinositols "GPI, GPIP (at full-scale) and $GPIP_2$". The [^{3}H]-GPI and GPIP peaks correspond to more than one [^{14}C]-glycerophosphate ester while incorporation of [^{14}C]-glycerol into $GPIP_2$ was insufficient for detection.

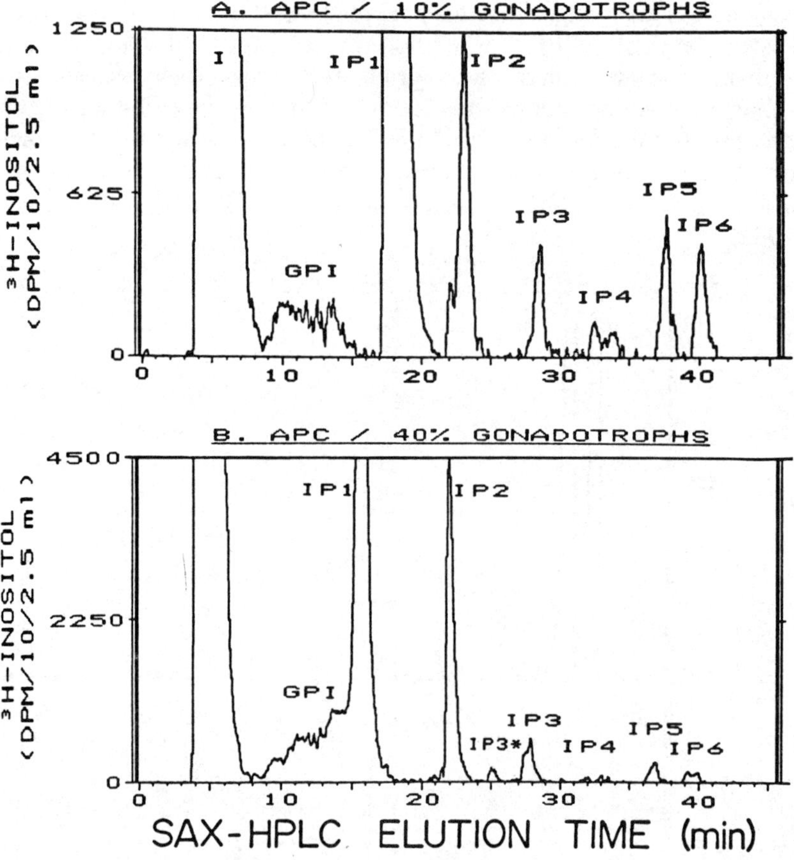

Figure 7. SAX-HPLC elution profile of inositol polyphosphates extracted from mixed anterior pituitary cells (upper panel A) and elutriated gonadotrophs (lower panel B). A: 6×10^6 anterior pituitary cells were cultured for three days in the presence of 3 μCi [^{3}H]-myo-inositol, washed, and stimulated with 10^{-7} *M* GnRH for 5 seconds. Elution was by rapid gradient program II with ammonium phosphate (0-0.6 *M* from 5-45 minutes). Excluding the [^{3}H]-inositol peak at 5 minutes, the 168K dpm were distributed into glycerophosphoinositol (5.4%); IP_1 (80.7%, shown cut-off at 1/11th of actual peak height); IP_2 at full-scale (6.5%); IP_3 (1.9%); IP_4 (2 peaks, 1.1%); IP_5 (2.3%) and IP_6 (2.1%). B: 1.5×10^6 gonadotroph-enriched APC were cultured for three days in the presence of 4 μCi [^{3}H]-myo-inositol, washing, and stimulation with 10^{-7} *M* GnRH for 15 seconds. Elution was by rapid gradient program II with ammonium phosphate (0-0.6 *M* from 5-45 minutes). Excluding the [^{3}H]-inositol peak at 5 minutes, the 221K dpm were distributed into glycerophosphoinositol plus Ins-1:2-cP (15.7%); IP_1 (67.7%, shown cut-off); IP_2 (12.2%, shown at full-scale); IP_3* (0.5%, probably Ins-1,3,4-P_3); Ins-1,4,5-P_3 (2.0%); IP_4 (0.4%); IP_5 (0.9%) and IP_6 (0.7%).

Further characterization of pituitary gonadotroph inositol phosphates was attempted with the use of a [^{3}H]-inositol plus [^{32}P]-phosphate double-labeling procedure. The crossover-corrected results of Figure 8 confirm the existence of both radioisotopes in IP species eluting after [^{3}H]-IP_3 and [^{32}P]-guanosine trisphosphate. The calculated phosphate/inositol ratios of the

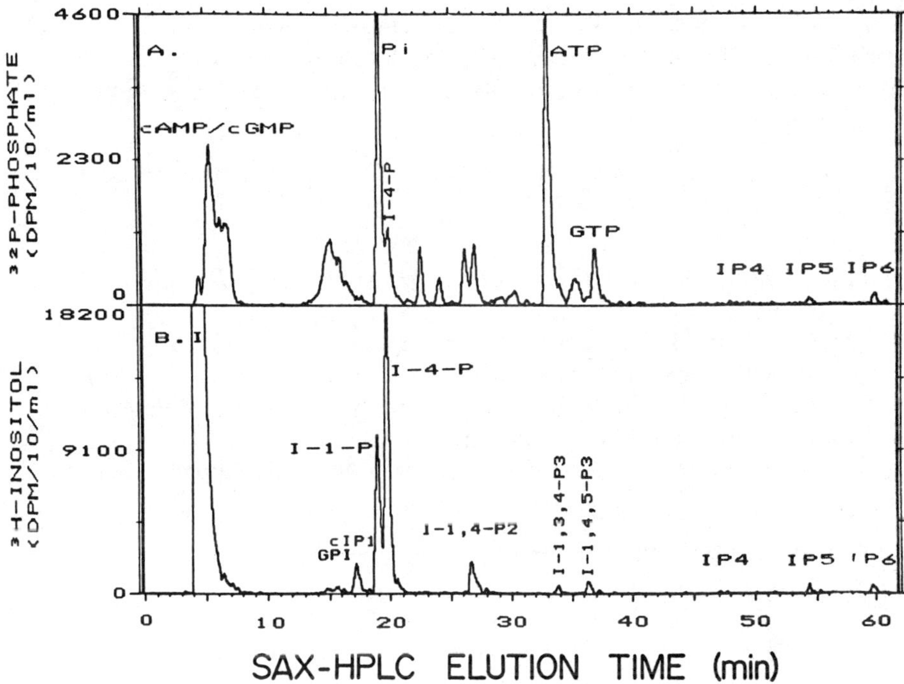

Figure 8. SAX-HPLC separation of [^{3}H]-inositol plus [^{32}P]-phosphate double-labeled polyphosphates from GnRH-stimulated, gonadotroph-enriched APC. One million cells on Cytodex-1 beads were cultured three days with 3 μCi [^{3}H]-myo-inositol and 3 μCi [^{32}P]-phosphate in "inositol-free" but phosphate-containing M199. Washed cells were then incubated in the presence of 10^{-7} *M* GnRH for 15 seconds followed by extraction of the phosphoinositols and nucleotides. Analysis was by rapid gradient program II (0-0.9 *M* ammonium phosphate between 5-65 minutes) with dual channel analysis of 146K dpm in [^{32}P]-labeled nucleotides plus phosphoinositols (panel A) and 182K dpm in [^{3}H]-inositol-labeled phosphoinositols (panel B). Linear regression analysis of the expected phosphate/inositol chemical ratios versus the observed [^{32}P]/[^{3}H] count ratios for Ins-4-P, Ins-1,4,5-P_3, IP_4, IP_5 and IP_6 showed significant correlation (r^2 = 0.96) and indicated an overall 14-fold greater incorporation of inositol versus phosphate into inositol monophosphates.

putative IP_4, IP_5 and IP_6, compared to the ratios found for the GPI derivatives of the parent phosphoinositides, substantiate the designation of these compounds as higher inositol phosphates.

An important additional feature of the IP analysis in Figure 8 is the resolution of isomers of IP_1, at 20 minutes and of IP_3 at 39-41 minutes. Although the chemical identification of the individual isomer peaks will be discussed below, it is worthy of note that this SAX-HPLC gradient system is unique in its ability to separate and identify isomers of all the principal IP intermediates from polyphosphoinositide metabolism (Morgan and Catt, 1987). The capability of directly comparing and independently quantifying [^{32}P]- labeled nucleotides or IPs and [^{3}H]-inositol-labeled IPs provides additional useful information; for example, the biologically important Ins-

1,4,5-P_3 coelutes with GTP while its inactive isomer Ins-1,3,4-P_3 elutes earlier with ATP in this system. The resolution of three isomers of IP_3 has also been achieved by decreasing the steepness of the ammonium phosphate gradient (Fig. 9).

IIIa. Weak anion exchange (WAX-) HPLC of inositol monophosphates w/phosphate

SM-A: HPLC-grade water.
SM-B: 12.5 mM HPLC ammonium phosphate to pH 3.35 with phosphoric acid.
SM-C: 0.5 *M* HPLC ammonium phosphate to pH 3.35 with phosphoric acid.
Column: Adsorbosphere 5μ NH_2; 0.46 x 25 cm steel (Alltech).
Mobile phase flow rate: 1.0 ml/min.
Sample injection: 0.2-2.0 ml of a neutral, low-salt, aqueous solution.

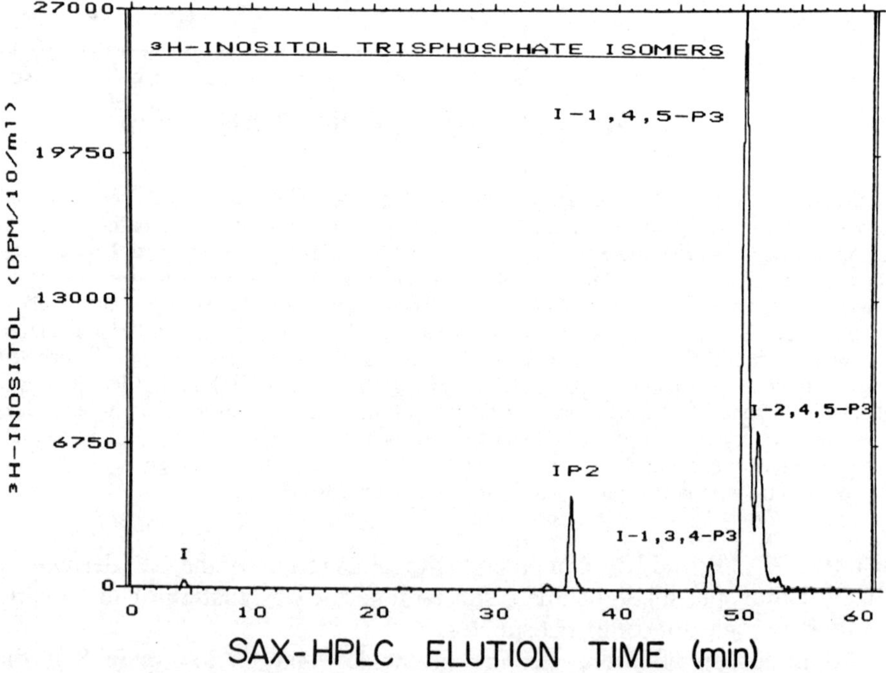

Figure 9. SAX-HPLC resolution of a crude [^{3}H]-inositol trisphosphate fraction prepared by Ca^{++}-activated phospholipase C hydrolysis of red blood cell [^{3}H]-PIP_2 (NEN). The major peak was separately found to coelute with authentic [^{3}H]-inositol-1,4,5-trisphosphate and the contaminating peak at 36.2 minutes coeluted with IP_2. The identities of Ins-2,4,5-P_3 and Ins-1,3,4-P_3 were based on their proportions and their elution time relative to known inositol polyphosphates and ATP. Total sample radioactivity was 261K dpm and, among the IP_3's, was distributed as 3.2% Ins-1,3,4-P_3, 75.8% Ins-1,4,5-P_3 and 21.0% Ins-2,4,5-P_3. Elution was by a slow gradient elution program II run from 0-0.4 *M* ammonium phosphate between 0-85 minutes.

"IP_1" Gradient Program IIIa: The standard program for resolution of inositol monophosphates and isomers (GPI, Ins-1:2-cP, Ins-1-P, Ins-2-P and Ins-4-P) began with a water-washed column at 1 ml/min SM-A. The initial gradient consisted of a linear increase to 100% SM-B at 1 ml/min between 0-50 minutes with the salt concentration changing at the rate of 0.25 mM/min. It should be noted that a "new" column may require substitution with a SM-B concentration up to 100 mM ammonium phosphate in order to achieve elution of inositol monophosphates similar to that presented here with a "used" column. This was followed by a linear gradient to 100% SM-C between 50-75 minutes, to "wash out" polyphosphorylated derivatives. The eluant was returned to 100% SM-A by 80 minutes and run at 1.5 ml/min for 15 minutes before returning to 1 ml/min SM-A prior to the next injection.

The importance of identifying isomers within the peaks of inositol monophosphates became evident when it was observed that only the second eluting of two IP_1 peaks on SAX-HPLC (Fig. 8) showed a significant increase upon gonadotroph stimulation by GnRH. One possibility was that an inositol-1:2-cyclic monophosphate produced by phospholipase C activation in gonadotrophs had yielded the corresponding IP_1 isomers upon hydrolysis, as shown in Figure 10. Weak anion exchange HPLC has previously been employed for the separation of inositol-1-monophosphate (Ins-1-P) from Ins-2-P (Hallcher and Sherman, 1980; Hokin-Neaverson and Sadeghian, 1984), but not with the mobile phase system described above. Good separation was achieved between Ins-1:2-cP, Ins-1-P and Ins-2-P and the expected high ratio of Ins-1-P to Ins-2-P resulted (Fig. 10), in contrast to the low ratio of IP_1 peak 1/peak 2 observed in stimulated cells (Fig. 8). It may also be noted that when cells were subjected to neutral extraction, only a very small peak of cyclic IP_1 was observed, and acid treatment yielded negligible Ins-2-P.

The occurrence of Ins-4-P in cells and a weak anion exchange method for its analysis was reported by Siess (1985). Synthetic Ins-4-P was prepared from [^{3}H]-PIP by strong base hydrolysis and analyzed by WAX-HPLC with dilute ammonium phosphate as described above (Fig. 11). Like Ins-2-P, Ins-4-P was retained approximately 5 minutes longer than authentic Ins-1-P on an aminopropyl column.

IIIb: Weak anion exchange (WAX-) HPLC of inositol monophosphates w/acetate

SM-A: HPLC-grade water.
SM-B: 50 mM HPLC ammonium acetate, to pH 4 with glacial acetic acid.
SM-C: 1.0 *M* HPLC ammonium acetate (pH 4).
Column: Adsorbosphere 5μ NH_2, 0.46 x 25 cm steel (Alltech).
Mobile phase flow rate: 1.0 ml/min.
Sample injection: 0.2-2.0 ml of a neutral, low-salt, aqueous solution.

"IP_1" Gradient Program IIIb: For comparison with program IIIa, another solvent system for inositol monophosphate separation was tested wherein

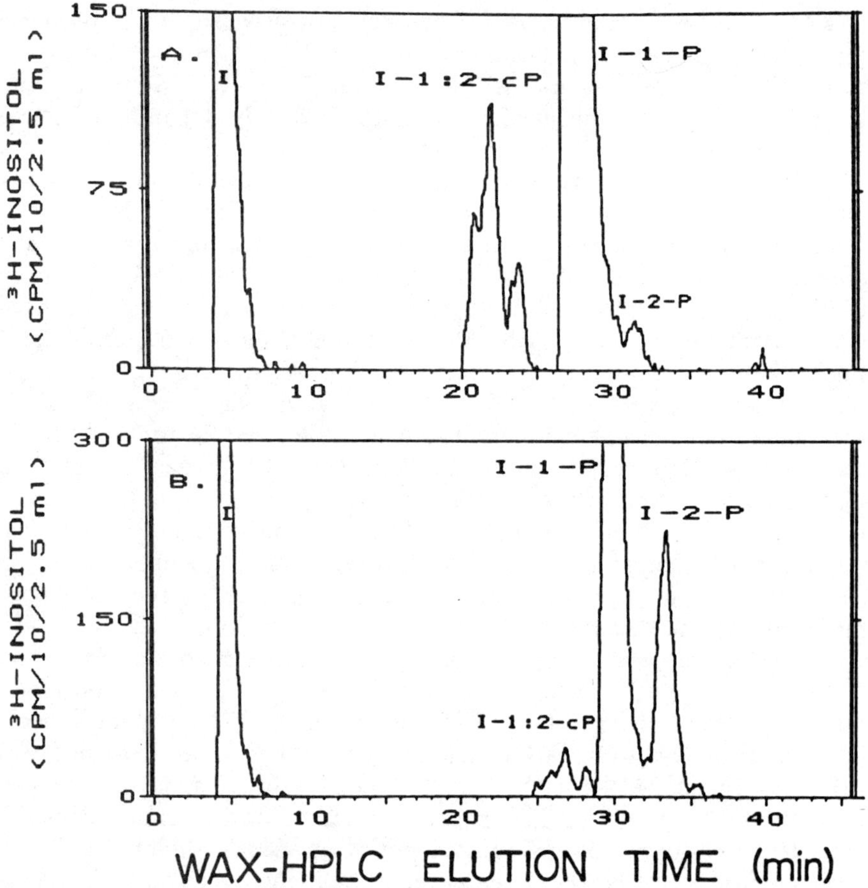

Figure 10. WAX-HPLC of the inositol monophosphates Ins-1:2-cP, Ins-1-P and Ins-2-P prepared by phospholipase C treatment at pH 5.5 of authentic [^{3}H]-inositol-labeled PI (NEN) followed by neutral, aqueous extraction only (panel A) or additional acid treatment (panel B). The major peak was identified by coelution with authentic [^{3}H]-Ins-1-P standard using the same elution program IIIa (0-12.5 mM ammonium phosphate over 0-50 minutes). Inositol-1:2-cyclic monophosphate and Ins-2-P are known products of these reactions and were identified by their relative elution times and precursor-product relationship under acid treatment. Total radioactivity in sample A was 30K cpm with 7% and 3% of the counts in Ins-1:2-cP and Ins-2-P respectively; radioactivity in sample B was 24K cpm with 1% and 11% of counts in Ins-1:2-cP and Ins-2-P respectively.

SM-B was replaced with 50 mM HPLC ammonium acetate brought to pH 4.0 with glacial acetic acid, and SM-C was replaced with 1.0 *M* ammonium acetate (pH 4). The elution gradient here was 100% SM-A to 100% SM-B from 0-50 minutes followed by a linear gradient to 100% SM-C between 50-100 minutes, all at a 1 ml/min flow rate. The column was washed with at least 25 ml water before the next injection.

To test whether the NH_2-column employed here was suitable for separation of all three inositol monophosphates, we chose a different mobile phase of ammonium acetate at much lower concentrations than used by Siess (1985),

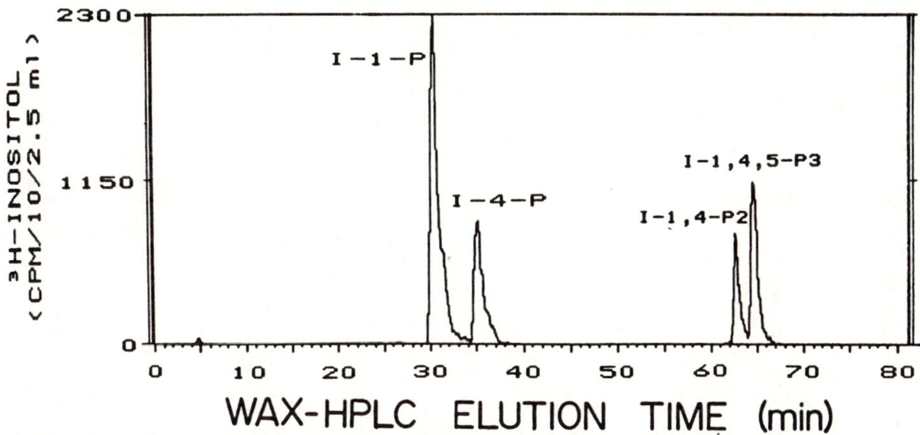

Figure 11. WAX-HPLC separation of [^{3}H]-Ins-4-P and [^{3}H] Ins-1,4-P_2, produced by strong base hydrolysis of [^{3}H]-PIP, from the internal standards [^{3}H]-Ins-1-P and [^{3}H]-Ins-1,4,5-P_3. The water-soluble PIP reaction products Ins-4-P and Ins-1,4-P_2 were detected as two distinct peaks of 10K and 6K cpm respectively in preliminary analyses. An equivalent sample with added Ins-1-P 25K cpm plus Ins-1,4,5-P_3 10K cpm analyzed here demonstrated that standard Ins-1-P eluted 5 minutes before the Ins-4-P reaction product during the shallow portion of gradient IIIa with ammonium phosphate on an aminopropyl-silica column. The Ins-1,4-P_2 byproduct was distinguished from IP_3 which eluted 2 minutes later during the steep gradient portion of the elution profile.

(gradient program IIIb). With the IP_1 standards prepared for the above mentioned analyses, we achieved separation of Ins-2-P and Ins-4-P by 5 and 7 minutes following Ins-1-P (Fig. 12). The same system was used to confirm that the second-eluting IP_1 peak on SAX-HPLC of pituitary cell metabolites (Fig.8) was Ins-4-P rather than Ins-2-P.

One final application of the WAX-HPLC gradient program IIIa was to establish that the Ins-4-P observed in cell extracts originated as a dephosphorylation product of Ins-1,4,5-P_3 — the immediate, bioactive product of polyphosphoinositide hydrolysis. Authentic, exogenous [^{3}H]-Ins-1,4,5-P_3 was incubated with the enzyme system of permeabilized pituitary cells and the products analyzed by WAX-HPLC in the absence (Fig. 13A) or presence of an Ins-1-P internal standard (Fig. 13B). The monophosphate product had the same elution characteristics as Ins-4-P, as distinct from Ins-1-P or Ins-2-P. Ins-5-P was not available for comparison, although it would likely be a metabolically labile candidate for cellular IP_1.

IVa. Reverse-phase (RP-) HPLC of arachidonate metabolites w/fraction collection

SM-A: 15% HPLC acetonitrile in water to pH 4.5 with trifluoroacetic acid.
SM-B: 90% HPLC acetonitrile in water to pH 4.5 with trifluoroacetic acid.
Column: Novapak 4μ C-18 reverse-phase, 0.8 x 10 cm cartridge (Waters).

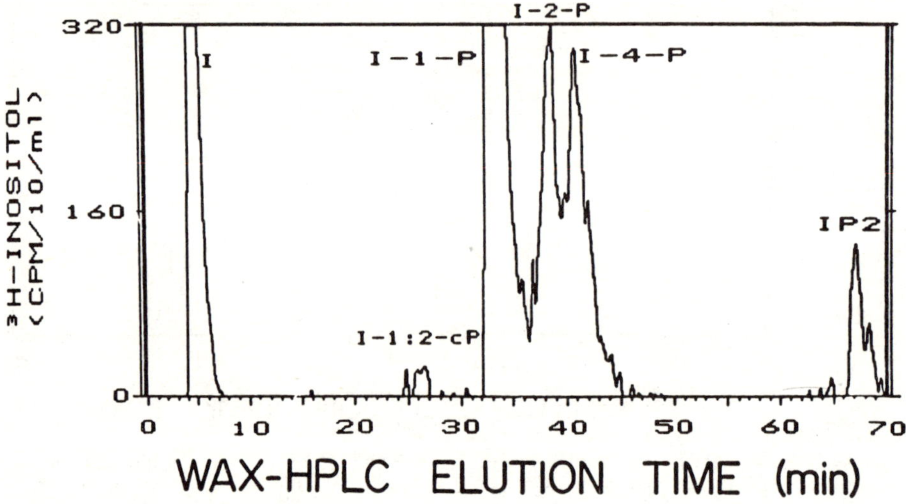

Figure 12. WAX-HPLC separation of three inositol monophosphate isomers synthesized from [^{3}H]-phosphoinositide standards. Ins-1-P and Ins-2-P (and free myo-inositol) were produced by phospholipase C treatment of PI followed by acid hydrolysis of the Ins-1:2-cP intermediate (see Fig. 10). Ins-4-P and Ins-1,4-P_2 were products of strong base hydrolysis of PIP (Fig. 11). The mobile phase consisted of dilute aqueous ammonium acetate, as described for gradient #IIIb. The combined sample comprised, in order of elution, 26% inositol, 0.5% Ins-1:2-cyclicP, 52% Ins-1-P, 9% Ins-2-P (at 38.6 minutes), 9% Ins-4-P (40.9 minutes) and 3% IP_2.

Mobile phase flow rate: 3.0 ml/min.
Sample injection: 200 microliter solution of fatty acids
in 50% SM-A + 50% SM-B.

"AA-M" Gradient Program IVa: Initial studies of arachidonate metabolites employed acetonitrile gradients with a radial compression cartridge system and a mobile phase flow rate of 3 ml/min, with 1-minute fraction collection and static liquid scintillation counting in Beckman Ready-Solv HP. The program began with a linear gradient from 90% SM-A plus 10% SM-B at injection to 62% SM-A plus 38% SM-B at 42 minutes. An isocratic run at 62% SM-A plus 38% SM-B between 42-93.5 minutes maintained the acetonitrile concentration at 43.5%. A second linear gradient to 40% SM-A plus 60% SM-B between 93.5-110 minutes was followed by an isocratic run at this 60% acetonitrile concentration for 40 minutes between 110-150 minutes. The final gradient to 100% SM-B between 150-155 minutes led to a 10-minute isocratic run at 100% SM-B, terminating at 165 minutes. The mobile phase was returned to initial conditions over 5 minutes and run at 90% SM-A plus 10% SM-B for 10 minutes prior to the next injection at 180 minutes.

Initial studies of arachidonic acid metabolism in pituitary gonadotrophs employed a C-18 reverse-phase cartridge system for metabolite analysis. In

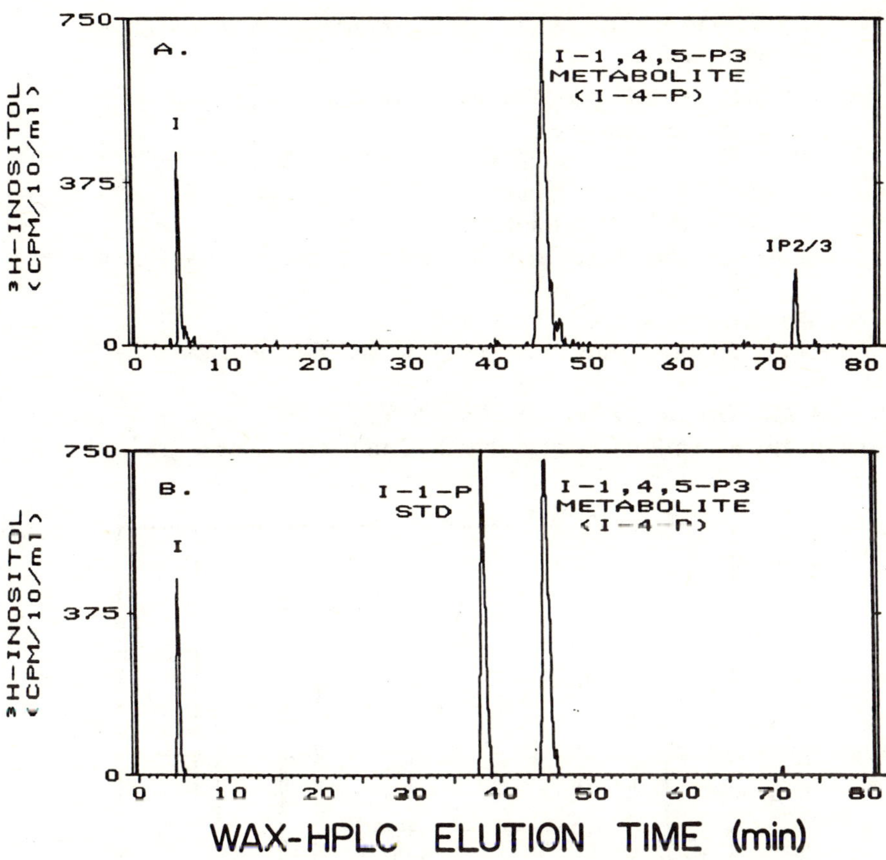

Figure 13. WAX-HPLC resolution of the pituitary cell phosphatase product of [^{3}H]-Ins-1,4,5-P_3 metabolism from an internal standard [^{3}H]-Ins-1-P. Five million mixed pituitary cells were permeabilized by 15 minutes incubation with saponin (200 μg/ml M199), and 50K dpm of authentic [^{3}H]-Ins-1,4,5-P_3 was added for an additional 60 minutes before termination and extraction. The IP_3 metabolite eluting here at 45 minutes coincided with both internal and external markers for synthetic Ins-4-P (panel A), as distinct from an Ins-1-P internal standard eluting 5.7 minutes earlier (panel B).

the absence of any previous comprehensive analysis of cellular AA metabolites we sought to identify cyclooxygenase, lipoxygenase ***and*** monooxygenase products of pituitary gonadotrophs in a single run (Fig. 14). The high flow rates involved with the radial compression system required manual collection and static scintillation counting of HPLC effluant fractions. Apart from the expected, major products of prostaglandin E_2 and 12-hydroxyeicosatetraenoic acid, we observed [^{3}H]-AA metabolite peaks eluting later than 12-HETE but before AA itself. In separate studies, those peaks were found to coelute with standard epoxyeicosatrienoic acid products of the hepatic microsomal cytochrome P_{450}-dependent monooxygenase enzyme system.

IVb. Reverse-phase (RP-) HPLC of arachidonate metabolites w/flow detection

SM-A: 0.1% HPLC phosphoric acid in water to pH 3.6 with triethylamine.
SM-B: 0.1% HPLC phosphoric acid in 0.5 mM oxalic acid (pH 5.4 triethylamine).
SM-C: 100% HPLC acetonitrile.
Column: Absorbosphere 3μ C-18 reverse-phase, 0.46 x 15 cm steel (Alltech).
Mobile phase flow rate: 1.0 ml/min.
Injection: 200 microliter solution of 50% SM-A + 50% SM-C.

"AA-M" Gradient Program IVb: 67% SM-A + 33% SM-C was run isocratically at 1 ml/min for the initial 22 minutes; a linear gradient to 67%

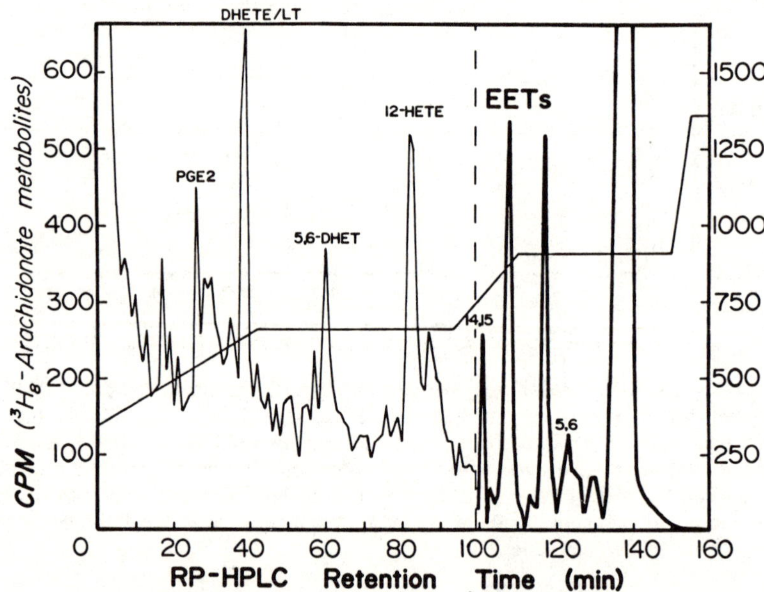

Figure 14. C18 RP-HPLC analysis of gonadotroph [^{3}H]-arachidonic acid metabolites. Elutriated gonadotrophs were prelabeled for one day with 5 μCi [^{3}H]-AA per three million cells. One sample of the washed cell suspension was incubated in the presence of 100 nM GnRH for three hours, after which lipids in the medium were extracted by C18 SEP-PAK. An acetonitrile gradient program IVa on a Novapak radial compression cartridge column was used to resolve cyclooxygenase, lipoxygenase and mono-oxygenase metabolites of endogenous AA. HPLC fractions (3 ml/min) were collected and radioactivity quantitated by static liquid scintillation counting. Four major, late-eluting peaks were identified as 14,15-, 11,12-, 8,9- and 5,6-epoxyeicosatrienoic acids (EETs) by comparison with [^{3}H]-AA peroxidized metabolites produced by a rat liver microsomal preparation. 5,6-dihydroxyeicosatrienoic acid (DHET), believed to be the hydrolase product of the labile 5,6-EET, eluted after cyclooxygenase products such as prostaglandin E_2 and between the dihydroxy and monohydroxy products of AA lipoxygenase.

SM-B + 33% SM-C was performed between 22-25 minutes, followed by a linear gradient to 62% SM-B + 38% SM-C from 25-55 minutes; an immediate switch to 62% SM-A + 38% SM-C at 55 minutes was continued isocratically to 60 minutes; a linear increase to 50% SM-A + 50% SM-C occurred between 60-85 minutes and this mixture was continued isocratically from 85-105 minutes; a linear rise to 35% SM-A + 65% SM-C from 105-120 minutes was maintained isocratic between 120-135 minutes; the final gradient was to 10% SM-A + 90% SM-C between 135-160 minutes and this was maintained for 10 minutes; after a linear change to the starting solvent mixture between 170 and 175 minutes, the mixture of 67% SM-A + 33% SM-C was maintained for 15 minutes until the next injection.

The need for a more efficient and economical RP-HPLC analysis of AA metabolites for routine metabolic studies mandated the conversion to a fine particle C-18 packing in a standard steel column with slower flow rates amenable to radiochemical analysis by flow scintillation counting. The new gradient program IVb was applied to the analysis of [^{3}H]-AA metabolites produced by prelabeled pituitary cells during short term stimulation by GnRH (Fig. 15). The result confirmed the predominance of EET

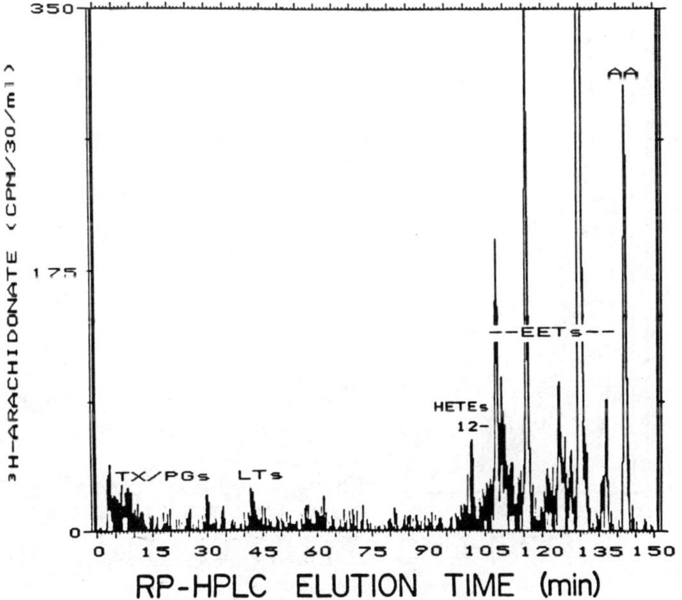

Figure 15. C18 RP-HPLC analysis of pituitary cell "EETs" following short-term stimulation by GnRH. [^{3}H]-AA-prelabeled cells were incubated with 100 nM GnRH for three hours, and the extracted metabolites were analyzed on a C18-3μ Adsorbosphere steel column with acetonitrile gradient IVb and liquid scintillation flow detection. Among at least five major peaks in the "EET" region, the largest peak at 130 minutes contained 52K dpm (66% of total) and may correspond to 8,9-EET (standards not rerun in this gradient system). Only traces of cyclooxygenase and lipoxygenase metabolites were observed in earlier regions of the chromatogram.

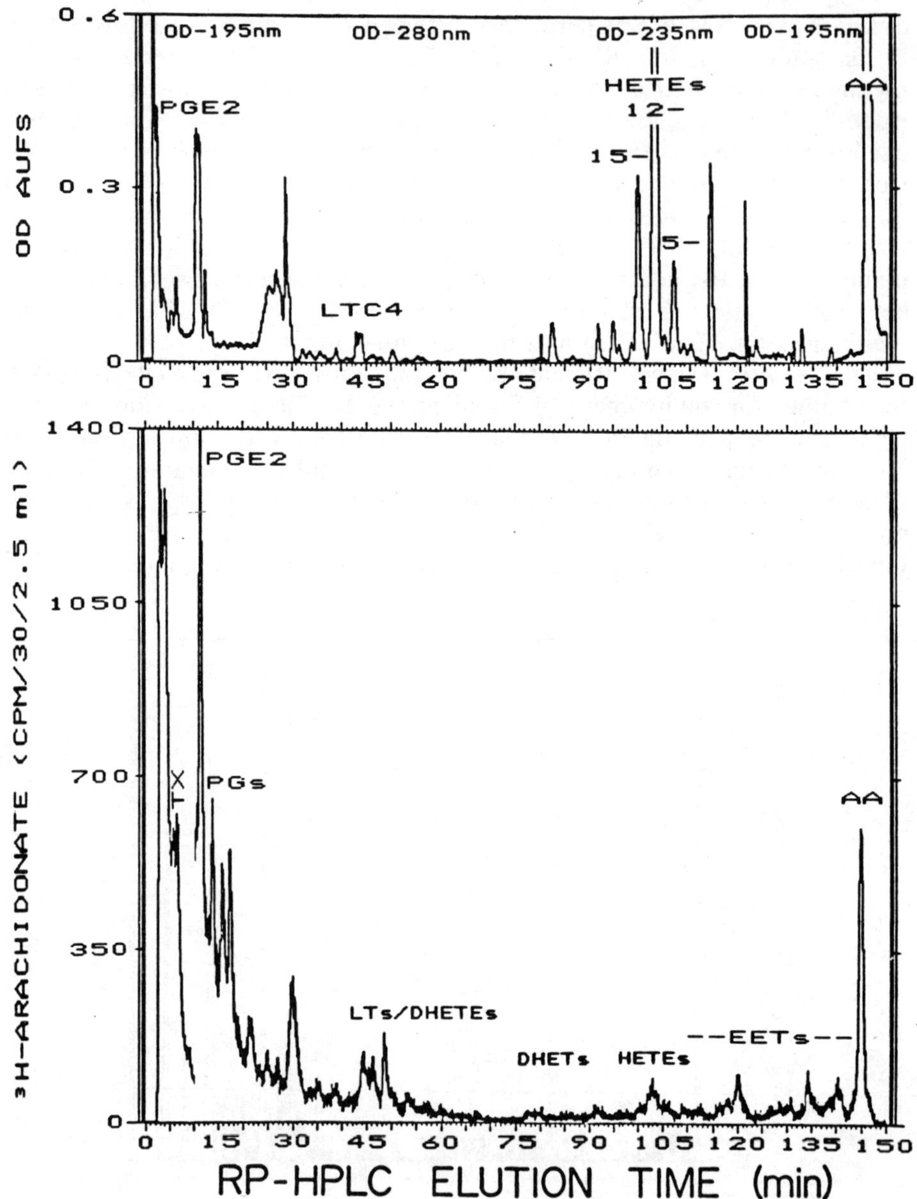

Figure 16. C18 RP-HPLC profiles of absorbance peaks from unlabeled standard AA metabolites (upper panel) and radiochemical flow detection of pituitary cell [^{3}H]-AA endogenous metabolites (lower panel). Elution was by acetonitrile gradient IVb on a C18-3μ steel column. PGE_2 at 11.5 minutes was the major metabolite (3.5% of the 1 million total cpm), following two days incubation at 37°C in the presence of exogenous [^{3}H]-AA without hormonal stimulation. Lipoxygenase products (DHETEs and HETEs) and monooxygenase products (DHETs and EETs) also accumulated to a lesser extent in the culture medium. Absorbance detection of unlabeled standards at appropriate wavelengths (upper panel) served to confirm the identity of certain radiolabeled metabolites (lower panel).

monooxygenase products over prostaglandins and HETEs that had previously been characterized. Resolution in this latter system compared to gradient program IVa above was achieved in equivalent time (ca. 145 minutes) but with one third the volume of mobile phase and liquid scintillation cocktail and under full automation.

The utility of gradient program IVb for complete AA metabolite analysis was demonstrated with UV absorbance detection of unlabeled standard metabolites and a different pattern of $[^3H]$-AA metabolites from pituitary cells following longer term incubation with the precursor (Fig. 16). Optimal wavelengths were chosen as an aid to the identification of specific metabolites concurrently measured by flow scintillation counting in another detection channel. The inclusion of $[^{14}C]$-AA metabolite standards has also been successful with programmed correction of channel crossover of low energy counts into the tritium channel containing sample peaks.

CONCLUSION

In summary, the collection of HPLC gradient programs presented here offer significantly improved analytical methodologies for the detection and isolation of a variety of products of cellular phosphoinositide metabolism. The use of flow scintillation detection for radiolabeled samples represents a quantum leap forward in facilitating automated routine analysis of these compounds with the additional advantages of economic and time savings for the laboratory.

Application of the tested methods to the study of GnRH action in pituitary gonadotrophs has yielded new information on the role of phosphoinositide metabolism in intracellular signaling. NP-HPLC and SAX-HPLC analysis of complex lipids has enabled us to identify and quantify changes in $[^3H]$-inositol-labeled PIP_2 and $[^{14}C]$-labeled DG immediately following phospholipase C activation by GnRH. The separation of Ins-4-P by SAX-HPLC and its identification as a product of Ins-1,4,5-P_3 metabolism by WAX-HPLC give a better understanding of the enzyme systems regulating IP_3 levels following cell activation. The double-label characterization of higher inositol phosphates in pituitary cells represents a new finding of possible relevance to IP metabolism in other mammalian cells. The production of potentially regulatory AA metabolites as a consequence of phosphoinositide metabolism can also be monitored with wider perspective and greater precision using the C-18 RP-HPLC modification presented here.

REFERENCES

Aitzetmuller, K. (1982). Recent progress in the high performance liquid chromatography of lipids. *Prog. Lipid Res.* 21, 171-193.

Bartlett, G.R. (1982). Isolation and assay of red-cell inositol polyphosphates. *Anal. Biochem.* 124, 425-431.

Berridge, M.J. (1985). The molecular basis of communication within the cell. *Sci. Amer. Oct.* 142-152.

Berridge, M.J. (1983). Rapid accumulation of inositol trisphosphate reveals that agonists hydrolyse polyphosphoinositides instead of phosphatidylinositol. *Biochem. J.* 212, 849-858.

Binder, H., Weber, P.C., Siess, W. (1985). Separation of inositol phosphates and glycerophosphoinositol phosphates by high-performance liquid chromatography. *Anal. Biochem.* 148, 220-227.

Brockerhoff, H. (1963). Breakdown of phospholipids in mild alkaline hydrolysis. *J. Lipid Res.* 4, 96-99.

Brockerhoff, H. and Ballou, C.E. (1961). The structure of the phosphoinositide complex of beef brain. *J. Biol. Chem.* 236, 1907-1911.

Clarke, N.G. and Dawson, R.M.C. (1981). Alkaline 0→N-transacylation. A new method for the quantitative deacylation of phospholipids. *Biochem. J.* 195, 301-306.

Creba, J.A., Downes, C.P., Hawkins, P.T., Brewster, G., Michell, R.H. and Kirk, C.J. (1983). Rapid breakdown of phosphatidylinositol 4-phosphate and phosphatidylinositol 4,5-bisphosphate in rat hepatocytes stimulated by vasopressin and other Ca^{2+} mobilizing hormones. *Biochem. J.* 212, 733-747.

Ellis, R.B., Galliard, T. and Hawthorne, J.N. (1963). Phosphoinositides. 5. The inositol lipids of ox brain. *Biochem. J.* 88, 125-131.

Frahn, J.L. and Mills, J.L. (1959). Paper ionophoresis of carbohydrates. I. Procedures and results for four electrolytes. *Aust. J. Chem.* 12, 65-89.

Grado, C. and Ballou, C.E. (1961). Myo-inositol phosphates obtained by alkaline hydrolysis of beef brain phosphoinositide. *J. Biol. Chem.* 236, 54- 60.

Hallcher, L.M. and Sherman, W.R. (1980). The effects of lithium ion and other agents on the activity of myo-inositol-1-phosphatase from bovine brain. *J. Biol. Chem.* 255, 10896-10901.

Hamilton, J.G. and Comai, K. (1984). Separation of neutral lipids and free fatty acids by high-performance liquid chromatography using low wavelength ultraviolet detection. *J. Lipid Res.* 25, 1142-1148.

Hamilton, J.G. and Karol, R.J. (1982). High performance liquid chromatography (HPLC) of arachidonic acid metabolites. *Prog. Lipid Res.* 21, 155-170.

Hendrickson, H.S. and Ballou, C.E. (1964). Ion exchange chromatography of intact brain phosphoinositides on diethylaminoethyl cellulose by gradient salt elution in a mixed solvent system. *J. Biol. Chem.* 239, 1369-1373.

Hokin-Neaverson, M. and Sadeghian, K. (1984). Lithium-induced accumulation of inositol-1-phosphate during cholecystokinin octapeptide- and acetylcholine-stimulated phosphatidylinositol breakdown in dispersed mouse pancreas acinar cells. *J. Biol. Chem.* 259, 4346-4352.

Hyde, C.L., Childs-Moriarty, G., Wahl, L.M., Naor, Z. and Catt, K.J. (1982). Preparation of gonadotroph-enriched populations from adult rat-anterior pituitary cells by centrifugal elutriation. *Endocrinology,* 111, 1421-1423.

Irvine, R.F., Letcher, A.J., Lander, D.J. and Downes, C.P. (1984). Inositol trisphosphates in carbachol-stimulated rat parotid glands. *Biochem., J.* 223, 237-243.

Isaacks, R.E., Kim, C.Y., Johnson, A.E., Goldman, P.H. and Harkness, D.R. (1982). Studies on avian erythrocyte metabolism. XII. The synthesis and degradation of inositol pentakis (dihydrogen) phosphate. *Poultry Sci.* 61, 2271-2281.

Jungalwala, F.B., Evans, J.E. and McCluer, R.H. (1984). Compositional and molecular species analysis of phospholipids by high performance liquid chromatography coupled with chemical ionization mass spectrometry. *J. Lipid Res.* 25, 738-749.

Kaduce, T.L., Norton, K.C. and Spector, A.A. (1983). A rapid, isocratic method for phospholipid separation by high-performance liquid chromatography. *J. Lipid Res.* 24, 1398-1403.

Kirk, C.J., Bone, E.A., Palmer, S. and Michell, R.H. (1984). The role of phosphatidylinositol-4,5-bisphosphate breakdown in cell-surface receptor activation. *J. Receptor Res.* 4, 489-504.

Morgan, R.O., Chang, J.P. and Catt, K.J. (1987). Novel aspects of gonadotropin-releasing hormone action on inositol polyphosphate metabolism in cultured pituitary gonadotrophs. *J. Biol. Chem.* 262, 1166-1171.

Osborne, D.J., Peters, B.J. and Meade, C.J. (1983). The separation of leukotrienes and hydroxyeicosatetraenoic acid metabolites of arachidonic acid by high performance liquid chromatography (HPLC). *Prostaglandins* 26, 817-832.

Patton, G.M., Fasulo, J.M. and Robins, S.J. (1982). Separation of phospholipids and individual molecular species of phospholipids by high- performance liquid chromatography. *J. Lipid Res.* 23, 190-196.

Saunders, R.D. and Horrocks, L.A. (1984). Simultaneous extraction and preparation for high-performance liquid chromatography of prostaglandins and phospholipids. *Anal. Biochem.* 143, 71-75.

Siess, W. (1985). Evidence for the formation of inositol-4-monophosphate in stimulated human platelets. *FEBS Lett.* 135, 151-156.

Tomlinson, R.V. and Ballou, C.E. (1961). Complete characterization of the myo-inositol polyphosphates from beef brain phosphoinositide. *J. Biol. Chem.* 236, 1902-1096.

Yandrasitz, J.R., Berry, G. and Segal, S. (1981). High-performance liquid chromatography of phospholipids with UV detection: Optimization of separations on silica. *J. Chromatog.* 225, 319-328.

Yandrasitz, J.R., Berry, G. and Segal, S. (1983). High performance liquid chromatography of phospholipids. Quantitation by phosphate analysis. *Anal. Biochem.* 135, 239-243.

Progress in HPLC, Vol. 3, pp. 57-78
Parvez et al. (Eds)

Quantitation of radioactive PTC derivatives of collagen crosslinks using reversed-phase chromatography and flow scintillation detection

ALBERT J. BANES and G. WILLIAM LINK
University of North Carolina, 253 Clinical Sciences,
Chapel Hill, North Carolina, USA

INTRODUCTION

Collagen crosslinks are the functional entities that impart structural integrity to collagen fibrils. Non-disulfide containing crosslinks may unite two, three or four collagen chains. The difunctional crosslink that occurs between chains of the same triple helix in the helical region is an intramolecular cross-link called aldol and is present in relatively small amount in collagen. The other difunctional crosslinks unite two chains between two triple helices (Gallup and Paz, 1975; Eyre, 1984). Crosslinks are composed of combinations of specific amino acids (lysine, hydroxylysine and sometimes histidine) in specific locations in the collagen triple helix. The majority of crosslinks unite the N or C terminal portion of a chain with that of the helical portion of another chain, in the case of difunctional crosslinks. In the case of tri or tetrafunctional crosslinks, three or four chains are covalently linked.

A collagen fibril is composed of two or three thousand collagen chains packed laterally. On the longitudinal axis, the molecules are packed head to tail in a quarter-stagger array where one 300 kd collagen molecule in a given row overlaps a second molecule below by 25% of its length. The crosslinks occur in this overlap region where participating amino acids of the N and C terminal portions interact to spontaneously form crosslinks with the chain below. Here is a specific example of how a difunctional crosslinks forms. A hydroxylysine in position 87 of the alpha-1-CB5 portion of a chain in a fibril is hydrolyzed at the epsilon amino group by lysyl oxidase, to form the aldehyde, hydroxyallysine (Kuboki et al., 1981). The epsilon amino group of hydroxylysine at position 17^c, in a second chain, in close proximity, reacts with hydroxyallysine 87 to form delta-hydroxy-alpha-aminoadipic acid delta-semialdehyde, the *in vivo* Schiff base crosslink. The Schiff base can be reduced with sodium borotritide resulting in the placement of tritium atoms on the carbon and nitrogen that participated in the Schiff base. However, the tritium atom on the nitrogen is labile and exchanges with non-radioactive protons. Therefore, the crosslink is onefold reduced with

NaB^3H_4. The reduced Schiff base is stable to acid hydrolysis, whereas the native compound is not.

Recently, Banes and coworkers have biochemically separated two compartments of collagen in bone: the nonmineralized and mineralized portions (Banes et al., 1983). These portions are separated by selective collagenase or trypsin hydrolysis of the nonmineralized surface collagen first, leaving the mineralized collagen behind. The mineralized collagen is subsequently demineralized with ethylenediaminetetraacetic acid at pH 8.0 then enzyme hydrolyzed (Banes et al., 1983). Banes and coworkers have also reported an unequal distribution of a trifunctional crosslink between the nonmineralized and mineralized collagen compartments (Banes et al., 1983).

The problems that have hampered this research are small sample size and lack of sensitivity in the detection of the crosslinks by amino acid analysis. Results of the current experiments indicate that the NaB^3H_4-reduced difunctional crosslinks, formed from the reaction of two hydroxylysines or a hydroxylysine and a lysine, can be derivatized with phenylisothiocyanate to form the phenylthiocarbamyl derivatives and separated in less than 20 minutes on a reversed-phase column. Sensitivity is enhanced by monitoring the column effluent with an on-line scintillation counter to detect radioactivity in crosslinks.

MATERIALS AND METHODS

Sample preparation

Samples from the nonmineralized (NMC) and mineralized (MC) compartments of bone, dentin, tendon, cartilage and bovine periodontal ligament were prepared according to a modification of the published method (Banes et al., 1983; 1985). Briefly, the method was as follows: a specimen was cleaned of extraneous soft tissue. For bone, the diaphysis alone was used and marrow discarded. Hard tissues were frozen at the temperature of liquid nitrogen and pulverized to 200-400 mesh powder using a SPEX freezer mill (SPEX Corp., Methuchen, NJ). The hard tissue samples (100 mg portions) were treated with ribonuclease and deoxyribonuclease to hydrolyze nucleic acids from disrupted cells. Samples were then treated with TPCK trypsin (1% wt./wt., Calciochem, Inc., Richmond, CA) in two, 30 minute treatments at 37°C to remove noncollagen protein and Type III collagen. Samples were then washed three times with 0.05 M PO_4 buffer (pH 7.2), to remove residual, hydrolyzed nucleic acids and noncollagen proteins. Samples were reduced with standardized NaB^3H_4 (ICN, Irvine, CA) (Banes et al., 1978). Samples were then reconstituted in 1 ml 0.02 M NH_4HCO_3 (pH 8.0), heated at 60°C for 20 minutes to denature collagen, cooled and hydrolyzed with half as much trypsin as was originally added. After hydrolysis, the samples were sedimented at 10,000 x g and the supernatant fluid collected, frozen and lyophilized in a vacuum drying oven. Samples were reconstituted in 1 ml of water and a portion removed for hydrolysis in Sequanol grade 6N HCl (Pierce Chemical Co., Rockford, IL) for one hour at 150°C after repeated N_2 flushing and evacuation using a Waters Pico-tag Work Station (Waters' Associates, Milford, MA) or a Refrigeration for Science automated sample hydrolyzer (Refrigeration for Science, Island Park, NY). Acid was volatilized in a Speed-Vac Concentrator with a chemical trap drying tower attached (Savant Inst., Hicksville, NY). Dried samples were stored at −20°C until prepared for derivatization for amino acid analysis.

For base hydrolysis, samples were hydrolyzed for 18 hours in 2N NaOH after repeated N_2 flushing and evacuation. Samples were neutralized by addition of an equal amount of HCl and desalted using a Sep-Pak column (Waters Associates, Milford, MA). Samples were then dried and stored at −20° until used.

PITC derivatization

Dried samples with approximately 100 μg to 1 mg of collagen were mixed with 1 μl of ethanol-water-triethylamine (2:2:1 by volume) per μg protein. Samples were redried twice with this solution, taking care to evacuate and hold the samples at less than 50 mTorr for 20 minutes. After redrying, samples were derivatized with 1 μl of a mixture of ethanol-triethylamine-water-phenylisothiocyanate (7:1:1:1) per 10 μg protein for 20 minutes at 25°C in the autohydrolysis instrument (modification of Heinrikson and Meridith, 1984; Jones et al., 1986). Samples were then evacuated to 50 mTorr and held for 20 minutes. Samples were reconstituted in 200 μl of 5 mM PO_4 buffer (pH 7.4), 5% acetonitrile, then sonicated and sedimented at 13,000 x g for five minutes. Radioactivity in a portion of the clear supernatant fluid was determined. The clear supernatant was used as the sample.

Amino acid analysis and crosslink determination

Amino acid analysis was performed using the Du Pont Reliance C18, 3 micron particle size, 60 angstrom pore size, analytical, reversed-phase column (40 x 6 mm, Reliance 3, C18 reversed-phase column, part #820668; 5 micron, 12.5 x 4 mm guard column, part #820674, with the column holder and column coupler system, Du Pont, Wilmington, DE). Two columns were used in series with a guard column prior to the analytical columns. These columns were chosen because (1) they could be screwed directly into each other and to the guard column, eliminating any dead space, and (2) because the first column could be discarded, the second column moved to the first position, and a new second column put in place, and (3) the Reliance column cartridges are inexpensive ($95.00 per cartridge).

The chromatography system used consisted of mobile phase A, 0.14 M sodium acetate (pH 5.7), with 700 μl triethylamine per liter; mobile phase B, CH_3CN, with a linear gradient of 7-36% B in 20 minutes at 1.5 ml/minute at 50°C, followed by 2.5 minutes at 95% 6 and 7.5 minutes of reequilibration with A.

An alternative system used involved dual C18 Du Pont 3 micron Reliance columns with a C18 guard column and a precolumn filter. Mobile phase A was 0.14 M sodium acetate with 700 microliters triethanolamine per liter, (pH 5.7). Mobile phase B was 100% acetonitrile. The gradient system included a flow rate of 1 ml per minute, 7% to 15% B in 15 minutes in a linear gradient, followed by 15% to 25% B in 30 minutes, followed by 25% to 100% B in 1 minute at 1.5 ml per minute with a hold for 4 minutes, after which the columns were reequilibrated to 7% B in 10 minutes. The total run time was 60 minutes. This procedure allowed more space in the chromatogram around lysine for the separation of the crosslinks.

Other elution systems were tested ranging from lower initial mobile phase B concentrations to using 60% acetonitrile as B and omission of triethylamine. The molarity of A was varied from 0.1 to 0.2 M sodium acetate. The pH of A was varied from 5.7 to 6.8. The current system was chosen because it gave the best separation of crosslinks and precursors. The guard column packed with a large particle size reversed-phase packing was used prior to the analytical column and gave no separation of the amino acids or crosslinks when used without the analytical column. A precolumn packed with a reversed-phase packing material was used between the pump and the injector.

Rheodyne column switching valves (Rheodyne Model 7030 switching valve, Cotati, CA) were used to select among the crosslink columns and a second series of Du Pont Reliance columns were used only for analytical amino acid separations.

A Waters chromatography system equipped with WISP, Model 840 system controller and integration system, dual 510 pumps, Model 440 detector at 254 nm, heater block and temperature programmer were used (Waters, a Division of Millipore Corp., Milford, MA). A Radiomatic FLO-ONE/Beta Model IC flow scintillation counter was used to determine radioactivity in the column effluent (Radiomatic Instruments and Chemicals Co. Inc., Tampa, FL 33611). In the scintillation counter, either a 500 or 2500 μl flow cell was used at a total flow rate of 5.0 ml/min (1 or 1.5 ml/min mobile phase; 3 or 3.5 ml/min Radiomatic FLO-SCINT III scintillation fluid). The 6 second update mode was used to avoid baseline distortion. A quench curve was stored to correct for increasing quenching throughout the separation. Data from the scintillation counter were printed in real time and stored on disc using dual Micromate drives in line with the Model IC computer, permitting subsequent reprocessing of data.

For amino acid determinations, and hydroxyproline analysis in particular, total sample loads in the 50 to 1000 picomole range were used with an amino acid standard injection in the same picomole range. For crosslink determinations, total mass loads up to 1 mg were used, and/or application of 10^4 to 10^6 dpm. Sample derivatization was scaled up to accommodate increased sample mass for crosslink samples.

Results

Figure 1a depicts the chemical structures of the aldehyde crosslink precursors as they exist *in vivo,* after chemical reduction and after reaction with

PITC DERIVATIZATION OF REDUCED CROSSLINK PRECURSORS HNL, DHNL

```
HOC=O              HOC=O              HOC=O  H
 HCNH2              HCNH2              HCNHCN
 HCH     RED        HCH    PITC        HCH  ‖
 HCH   ------>      HCH   ------>      HCH  S
 HCH   NaB3H4       HCH                HCH
  C=O               HCH3               HCH3
 H                   OH                 OH
 AK                 HNL                PTC-HNL

HOC=O              HOC=O              HOC=O  H
 HCNH2              HCNH2              HCNHCN
 HCH     RED        HCH    PITC        HCH  ‖
 HCH   ------>      HCH   ------>      HCH  S
HOCH   NaB3H4      HOCH               HOCH
  C=O               HCH3               HCH3
 H                   OH                 OH
AOK                 DHNL               PTC-DHNL
```

Figure 1a. Structures of the aldehydes of lysine (K) and hydroxylysine (OK), allysine (AK) and hydroxyallysine (AOK) respectively; the sodium borohydride-reduced compounds, hydroxynorleucine (HNL) and dihydroxynorleucine (DHNL); and the phenylthiocarbamyl derivatives of the latter, PTC-HNL and PTC-DHNL.

phenylisothiocyanate (PITC). One tritium atom is placed on the epsilon carbon of either allysine (AK) or hydroxyallysine (AOK) after reduction of the aldehyde to an alcohol with tritiated sodium borohydride (NaB^3H_4). Reaction of the reduced precursors at the alpha amino group with PITC yields the phenylthiocarbamyl derivatives of the precursors. One mole of PITC reacts per mole of precursor (Fig. 1a).

Figure 1b illustrates the structures of the crosslinks of lysine and hydroxylysine as they exist *in vivo*: the crosslink formed by reaction of OK and AOK is delta^{6,7}-hydroxy-delta-DHLNL; the crosslink formed from K and AOK or AK and OK is delta^6-hydroxy-delta-HLNL. Upon reduction with NaB^3H_4, the Schiff base crosslink (between the aldehyde precursor and the epsilon amino group of either K or OK) is reduced resulting in the placement of one tritium atom on the epsilon carbon and one on the nitrogen of the Schiff base. Therefore, two tritium atoms react with each crosslink during reduction. However, the tritium atom on the nitrogen is labile and freely exchanges with the nonradioactive protons in the medium. The result is that the crosslink is onefold reduced. During derivatization, three moles of PITC react per mole of crosslink: one mole each at the two alpha amino groups, and one at the secondary amine at the reduced crosslink region (Fig. 1b). When the PTC-crosslink is formed, the hydrogen atom is displaced from the secondary amine as PITC reacts to form the PTC derivative.

Data in Figure 2 depict the reducible crosslink pattern for adult, human, dentin mineralized collagen. The major reducible crosslink in dentin is

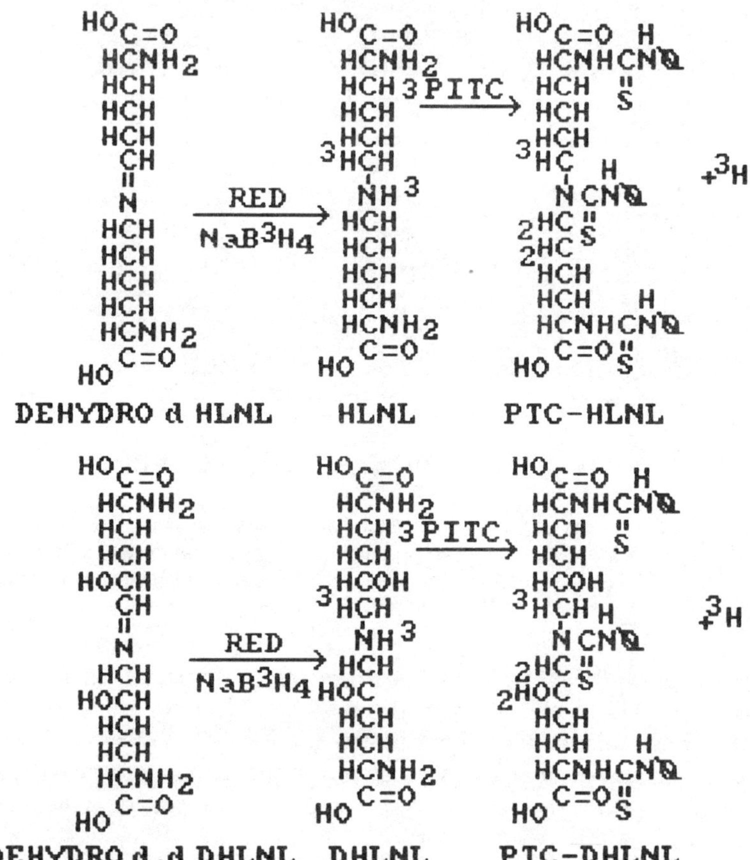

Figure 1b. In vivo structures of the crosslink composed of OK and AOK, delta^{6,7}-hydroxy-delta, delta-DHLNL, and the crosslink composed of K and AOK or OK and AK, delta^{6}-hydroxy-delta-HLNL. A Schiff base is the functional group of these divalent crosslinks. Reduction with sodium borohydride adds tritium to the Schiff base double bond resulting in one tritium atom on the epsilon carbon and one on the nitrogen. The tritium on the nitrogen is labile and exchanges with free protons in the medium. The reduced crosslink that forms after reaction of OK and AOK is dihydroxylysinonorleucine (DHLNL); the reduced crosslink that forms from K and AOK or OK and AK is hydroxylysinonorleucine (HLNL). Reaction of the reduced crosslinks with PITC requires three moles of PITC per mole of crosslink (Fig. 1b).

dihydroxylysinonorleucine DHLNL (Mechanic, 1974). Only the DHLNL crosslink (2) and its precursor, dihydroxynorleucine (DHNL) are present in this sample. Data in Figure 2 indicate the elution position of the crosslink, DHLNL and the precursor, DHNL in this tissue (Fig. 2). The first prominent radioactive peak that eluted consisted of $^{3}H_2O$ and NaB^3H_4. These peaks could be eliminated by desalting on a Sep-Pak column (Waters' Associates, Milford, MA), or other suitable reversed-phase packing material.

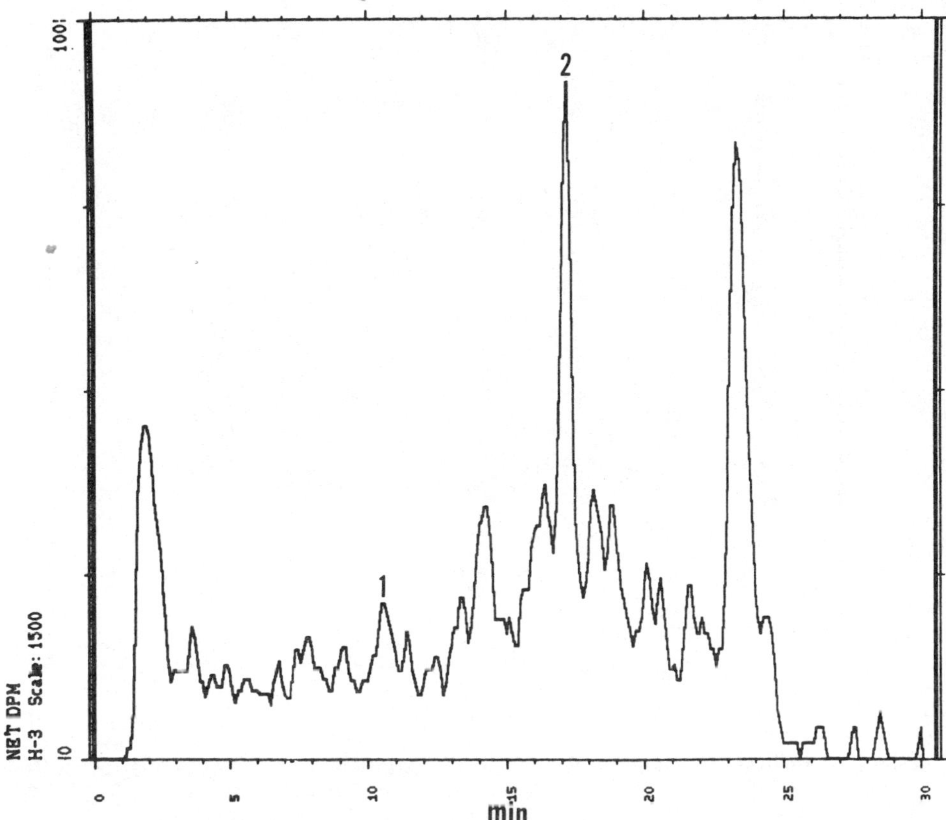

Figure 2. Radiochromatogram of adult, NaB^3H_4 reduced, PITC-derivatized, human dentin mineralized collagen. The principal reduced crosslink is dihydroxylysinonorleucine (DHLNL) that forms from reaction of AOK and OK. The crosslink eluted with K on a Du Pont Reliance reversed-phase column (see methods). The precursor, DHNL, eluted before hydroxynorleucine in this system. The radioactive components that eluted from 1 to 1.5 minutes were ^3H_2O and residual NaB^3H_4. The peak that eluted at 22.5 to 25 minutes was an artifact peak associated with compounds that eluted with 100% B (60% CH_3CN). This peak appeared with the mobile phase interface of 100% B and 7% B during reequilibration.

The prominent radioactive peak at the end of the chromatogram (22.5 to 25 minutes) was an artifact peak coinciding with PITC reagent peaks and the eluent shift from 100% B to 7% B during reequilibration.

Data in Figure 3a depict the radiochromatogram of collagen crosslinks from adult, avian, flexor tendon from the tibiotarsus region of the leg. Both dihydroxylysinonorleucine (DHLNL, peak 4) and hydroxylysinonorleucine (HLNL, peak 5) are present as the reduced crosslinks. The position beneath the numeral 3 is the approximate elution position for lysine.

Data in Figure 3b depict the radiochromatogram of collagen crosslinks of adult, rat tail tendon. The principal reduced crosslink in this sample is HLNL (peak 3) (Housley and Tanzer, 1981).

The radiochromatogram in Figure 3c indicates the presence of three

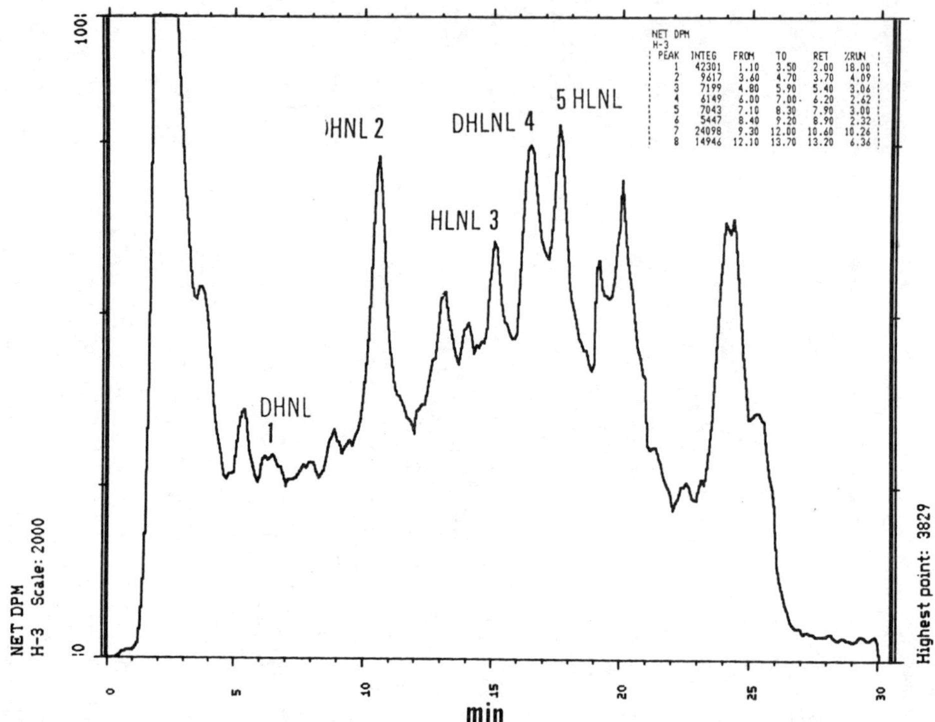

Figure 3a. Radiochromatogram of reduced, PITC-derivatized, avian flexor tendon collagen. The crosslinks HLNL (3) and DHLNL (4) elute with and just after K respectively. The radioactive precursors, DHNL (1) and HNL (2) eluted just before and after I and L. *Bone fide* HNL eluted as a split peak with detection at 254 nm.

crosslinks: DHLNL, eluting near lysine, HLNL, eluting post-DHLNL, and the tetrafunctional crosslink, histidinohydroxymerodesmosine (HHMD) eluting after HLNL. This pattern was confirmed with ion exhange separation of the crosslinks from the same sample.

Figure 4 represents the radiochromatogram of avian epiphyseal cartilage. A major crosslink in this sample is also DHLNL (peak 2) (Eyre and Oguchi, 1980).

Data in Figures 5a and 5b indicate the elution positions and relative amounts of the two major collagen crosslinks in normal and osteoblastoma avian bone, DHLNL and HLNL (Banes et al., 1978; 1983). Figure 5a represents the crosslink pattern in the nonmineralized collagen compartment (NMC) of normal bone. The NMC, or surface collagen that is hydrolyzed by trypsin after heat denaturation has approximately equal amounts of DHLNL and HLNL (Fig. 5a). The ratio of the two crosslinks is 1.1 (Table 1).

Data in Figure 5b indicate the crosslink pattern in the mineralized compartment of normal avian bone (MC, Fig. 5b). Although the DHLNL/HLNL ratio (peaks 3/4) is 1.02 for the MC, the total amount of the crosslinks is less than that in the NMC (Table 1).

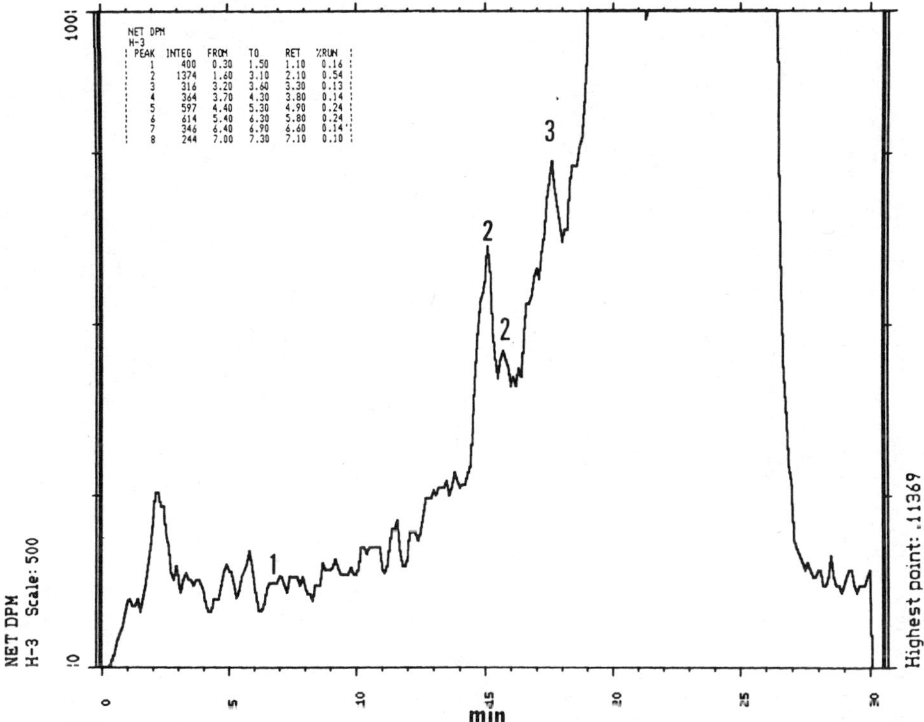

Figure 3b. Radiochromatogram of reduced, derivatized adult, rat tail tendon collagen. The peak that eluted after K is HLNL, the characteristic crosslink of this tissue.

Data in Figures 6a and 6b show the crosslink patterns for the NMC and MC of osteoblastoma bone, a disease induced by a virus (Mycloblastosis-associated virus-2(Osteo), (Banes and Smith, 1977). The ratio for the crosslinks (DHLNL/HLNL, peaks 3/4) in both NMC and MC compartments is 1.1 and 0.837 respectively (Table 1). However, the osteoblastoma bone sample has more total crosslinks than the normal sample, and has more crosslinks in the NMC compartment than in the MC compartment (Table 1).

Data in Figures 7a and 7b indicate radiochromatogram patterns for avian, normal and osteoblastoma bone MC collagen that had been reduced with NaB^3H_4, then hydrolyzed in 2N NaOH. Peaks 1 and 2 are DHLNL and HLNL respectively.

The chromatogram in Figure 8a indicates a typical analytical separation of phenylthiocarbamyl derivatives of amino acids using the same gradient program used to separate the collagen crosslinks and precursors (Fig. 8a). Hydroxyproline (OP) is well resolved from glutamic acid and serine and readily quantitated.

The chromatogram in Figure 8b represents the absorbance profile at 254 nm of a typical crosslink sample separation. The peak heights of the majority of the amino acids are far greater than those of the crosslinks which elute in the region of lysine. A reagent peak elutes between DHLNL and HLNL.

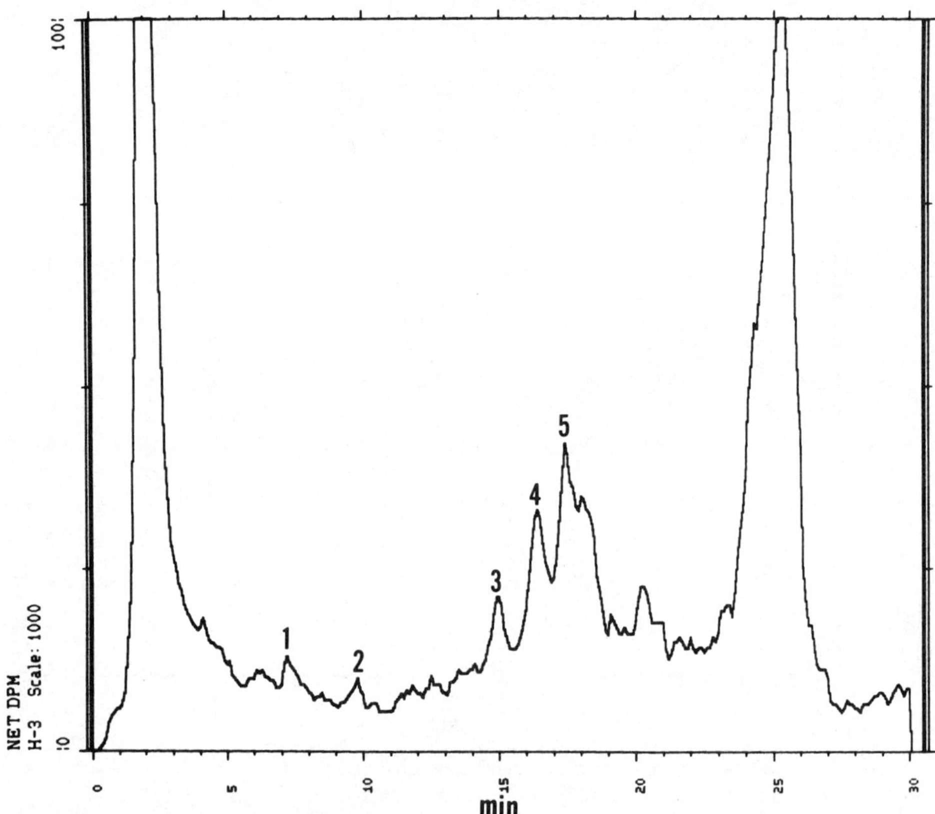

Figure 3c. Radiochromatogram of reduced, PITC-derivatized bovine periodontal ligament. Crosslink precursors and crosslinks are labeled (1), DHNL; (2), HNL, (3), unknown; (4), DHLNL; (5), HLNL, histidinohydroxymerodesmosine, (HHMD eluted after HLNL). The presence and amounts of these crosslinks were confirmed by chromatography in an ion exchange system.

The precursors, DHNL and HNL elute between 8 and 11.5 minutes. These times are greater than the actual chromatographic times due to the use of the 2.5 ml mixing cell in the Radiomatic flow scintillation counter.

Data in Table 1 list the crosslinks and precursor specific activities of three month-old normal and osteoblastoma avian bone (tibiotarsus) as radioactivity in dpm per micromole hydroxyproline. The ratios of the crosslinks and precursors are also given.

DISCUSSION

Past methodology and equipment

Methodology in collagen chemistry has been positively affected by the utilization of reversed-phase packings for the separation of collagen chains, tryptic and cyanogen bromide peptides, as well as crosslinks and amino

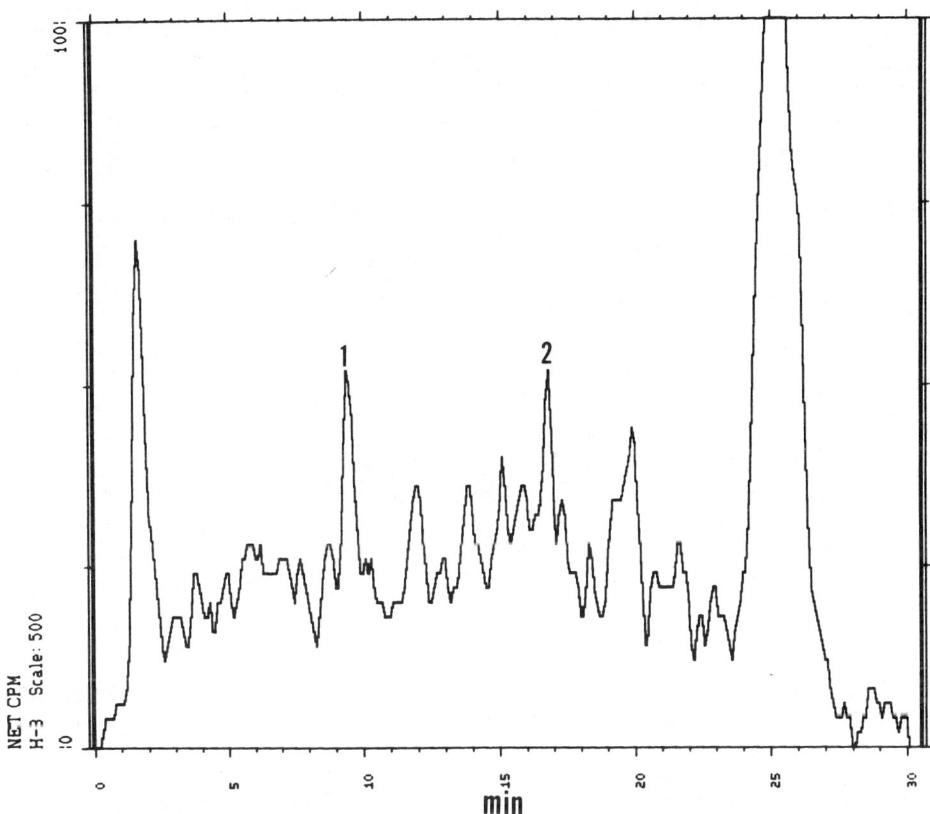

Figure 4. The radiochromatogram indicates that DHLNL was a major crosslink (2) in a sample of avian, epiphyseal cartilage.

acids (Van der Rest et al., 1980; Fallon et al., 1981; Banes et al., 1983; Reiser and Last, 1983; Banes et al., 1985; Smolenski, 1984; Eyre et al., 1984; Link et al., 1985). Quantities of sample required for detection of crosslink peptides have declined from 100 mg to less than 100 μg in the past five years with improved methodology (Yamauchi et al., 1981; Banes et al., 1983, 1985). Likewise, the time required to effect a peptide separation has declined from three days with a three meter G-50 superfine column, to one to two hours using an 8 cm to 25 cm reversed-phase column. Sample size and separation time for collagen crosslinks have also declined from mgs to μgs and hours to less than one hour respectively using high performance ion exchange or reversed-phase chromatography (Smolenski et al., 1984; Reiser and Last, 1983; Eyre et al., 1984).

Another improvement which has increased detectability of collagen crosslinks is the on-line, flow scintillation counter. The prototype of this instrument has been in use in a limited number of laboratories as a specially designed instrument for a number of years (Mechanic, 1974). This prototype accepted column effluent and mixed it with scintillation fluid (toluene,

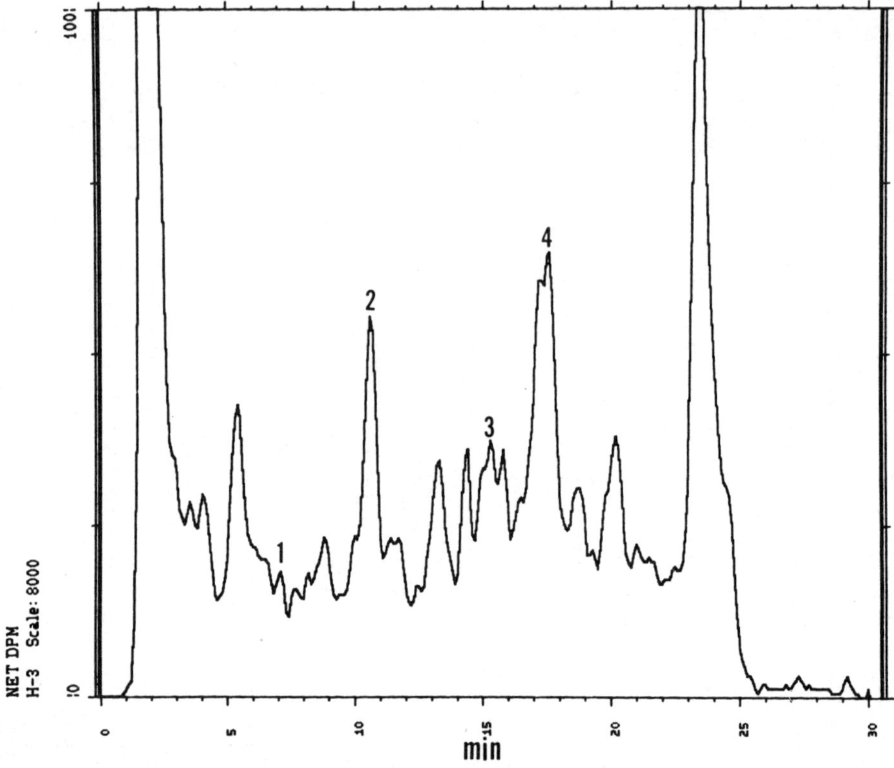

Figure 5a. Radiochromatogram of a sample of nonmineralized collagen (NMC) from a normal, 3 month-old chick tibiotarsus diaphysis. There was less DHNL (2) precursor than HNL (1). Both DHLNL (peak 3) and HLNL (peak 4) were prominent.

Beckman Biosolve BBS-3 and butyl PBD fluor- this cocktail does not gel at high salt or aqueous conditions) dynamically in a sample mixing vial. The column effluent and scintillation fluid mix was then aspirated into a specially designed counting vial in a Beta-Mate scintillation counter (Beckman Inst., CA). The counting vial had a tube for the sample input and a second tube that aspirated the sample to waste. A valving system regulated the flow of the sample to the counting vial or to waste. The entire operation functioned on a 60 second cycle: a sample was collected from the column for 60 seconds, then aspirated into the counting vial and counted for 30 seconds. After counting, the sample was aspirated to waste and the next sample was aspirated into the counting vial, etc. Data were printed out in counts per minute with a printer interfaced to the beta counter. This system worked well for lengthy separations, but would not be suitable for peaks that elute with a bandwidth of less than one minute, or for multiple components that elute near the same k'.

Development of more compact, reliable flow scintillation counters has increased the accessibility of this useful technique to the general scientific population. Interfacing a computer to the detector has permitted raw data

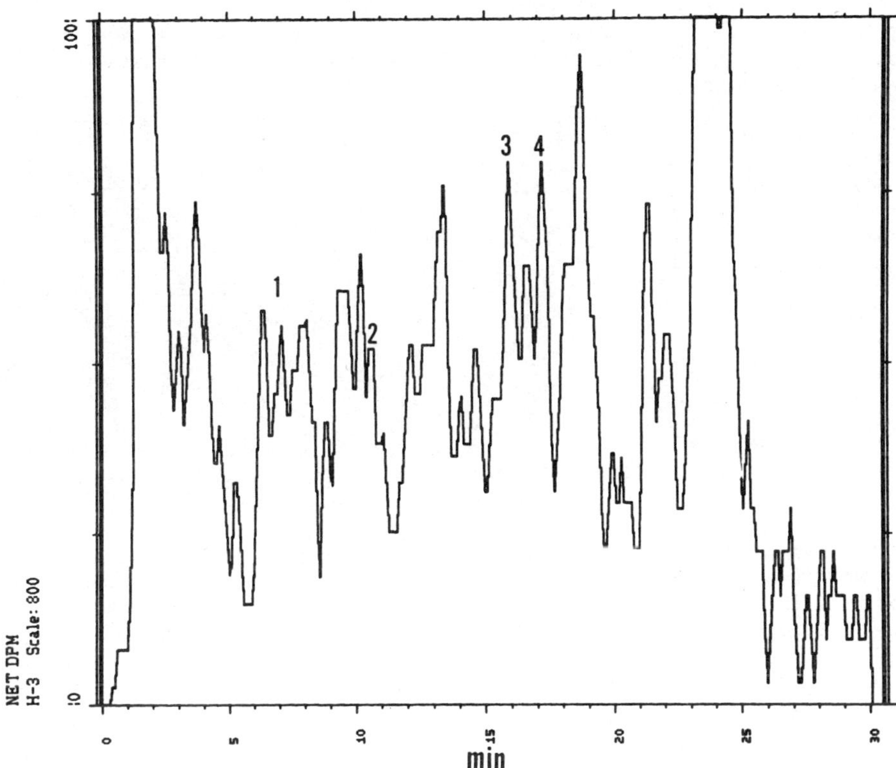

Figure 5b. Radiochromatogram of the mineralized compartment (MC) from the same sample as in Figure 5a. The amount of radioactivity was low in this sample and the use of the 500 μl flow cell in the scintillation counter yielded a poor baseline. The principal crosslinks and precursors were still detectable.

collection, storage and retrieval for further data computation. Some currently available commercial flow detectors have these capabilities (Kessler, 1982).

Flow scintillation counter conditions

The Radiomatic FLO-ONE/Beta Model IC flow radiation detector with dual disc drives for main program loading, access and data storage and handling, has greatly improved the efficiency of testing different chromatographic methods for the crosslink separations. Raw data were stored and reprocessed in different modes such as counts per minute, disintegrations per minute or as cpm or dpm in integrated peaks. The reprocess mode was essential to bring peaks on scale and most importantly, to verify that integrated peaks really accounted for all the radioactivity attributed to them by the integration program. It was also essential to correct for efficiency of counting under all the different chromatographic conditions tested. This was performed by using the internal program that allows one to calculate quenching based on the type of scintillation fluid used, the mobile phases and the

Table 1. Specific activity as dpm per μmole hydroxyproline of reducible precursors and crosslinks in collagen from 3 month-old normal and osteoblastoma chick bone nonmineralized and mineralized collagens.

Bone	1 DHNL	2 HNL	3 2/1	4 DHLNL	5 HLNL	6 5/4	7 4+5	8 4+5 N+M
Normal								
NMC	9156	3010	3.04	15797	7822	2.02	23619	26375
MC	3365	524	6.42	1343	1414	0.95	2756	—
AO								
NMC	5955	4740	1.26	20357	15631	1.30	35988	45985
MC	2898	846	3.43	5926	4071	1.46	9997	—

Precursors and crosslinks were separated on a Du Pont Reliance reversed-phase column as the NaB^3H_4-reduced, PITC-derivatized compounds.

Nonmineralized (NMC) and mineralized (MC) refer to the collagen compartments found in bone.

DHNL and HNL are the precursors dihydroxynorleucine and hydroxynorleucine respectively. DHLNL and HLNL are the crosslinks dihydroxylysinonorleucine and hydroxylysinonorleucine respectively. Values in column 3 represent the ratio of the precursors enumerated in columns 2 and 1. Values in column 6 represent the ratio of the crosslinks enumerated in columns 4 and 5. Values in column 7 represent the sum total of crosslinks in the NMC or MC compartments of a sample. Values in column 8 represent the total number of crosslinks in both the NMC and MC of a sample.

flow rates of the scintillation fluid and the column effluent. Curves could be constructed automatically by using a scintillation cocktail with a known amount of tritium in it, or by editing a curve and adding one's own counting efficiency values. For our experiments, the scintillation fluid used was either a commercial brand (FLO-SCINT III, Radiomatic, Tampa, FL) or a cocktail based on pseudocumene, BBS-3 and butyl-PBD. The latter had only 50% of the efficiency of the commercial brand, therefore the commercial brand was used. It was important to maintain fluidity in the scintillation fluid-sample mix, particularly at high flow rates.

We also found that the 500 μl flow cell gave baseline noise problems at high sensitivity (full scale 100 dpm to 500 dpm). If one had sufficient radioactivity to have thousands of counts per peak or few peaks, then the 500 μl flow cell was adequate. We preferred this flow cell with clean samples because we had better resolution of closely eluting components. We routinely used the 2500 μl flow cell however, because baseline noise was less of a problem with samples that did not contain highly radioactive components. Even components that eluted within less than one minute of each other, and were not resolved to baseline in the chromatographic system, could be distinguished and quantitated using the Radiomatic FLO-ONE/Beta Model IC unit.

Chromatography conditions

During the course of chromatographic methods development for these experiments, different gradient conditions were used to separate amino acids

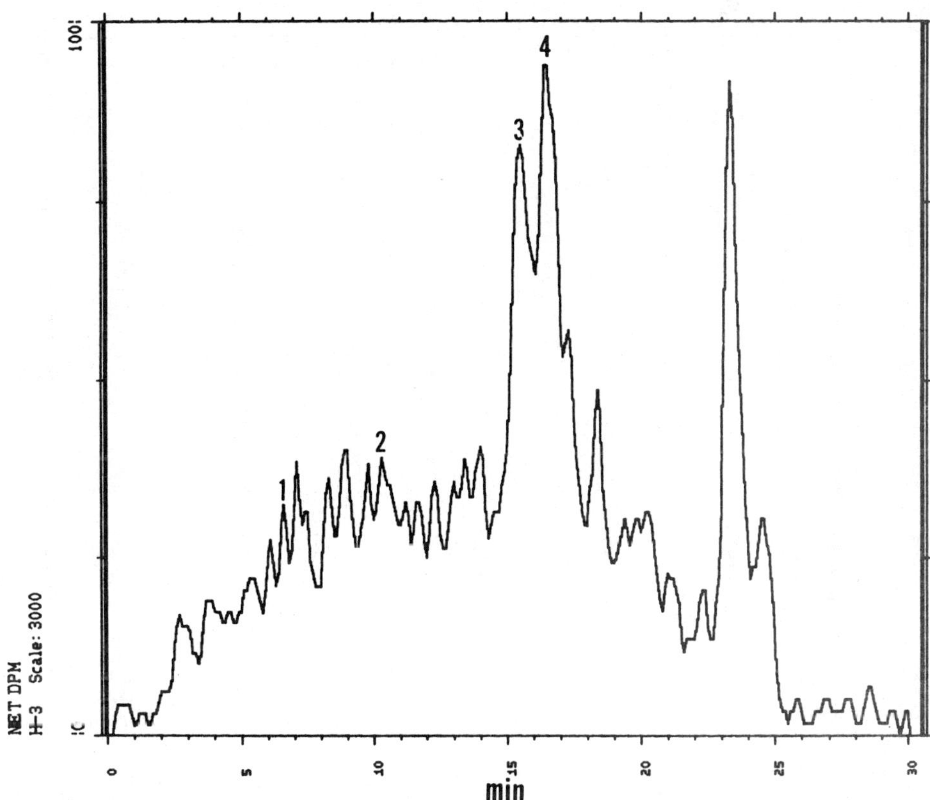

Figure 6a. Radiochromatogram of the NMC of a sample of osteoblastoma bone from a 3 month-old chick that was infected with MAV-2(O) *in ovo*. The reduced crosslinks (3, DHLNL; 4, HLNL) were clearly evident, although not resolved to baseline due to the great specific activity of the compounds. The precursors were also present (1,2).

and crosslinks. Variation in the molarity (0.1 M to 0.2 M sodium acetate) and pH (between 5.7 and 6.8) of mobile phase A altered the elution pattern of the PTC amino acids. These alterations were subtle elution profile changes compared to those incurred by increasing the mobile phase B strength. Low initial B concentrations (1 to 2 percent), retarded elution of the PTC amino acids. Good separation occurred with the acidic amino acids and the rest of the front third of the eluting components. R,T,MSO merged to form peaks with shoulders (Fig. 8b). Moreover, the PITC reagent peaks post-lysine merged with the crosslink region. Increasing the time over which the 7% to 36% B gradient occurred did not yield a better separation. Changing the gradient parameters to 7% to 25% B, using dual Reliance C3 columns and extending the run time to 50 minutes, increased the resolution of DHLNL and HLNL (see methods section). Increasing the flow rate from 1 ml/min to 1.5 ml/min increased the plate number by approximately 30% and did yield a greater peak capacity for the Reliance columns as predicted by an experimental computer model (Stadalius et al., 1985a,b).

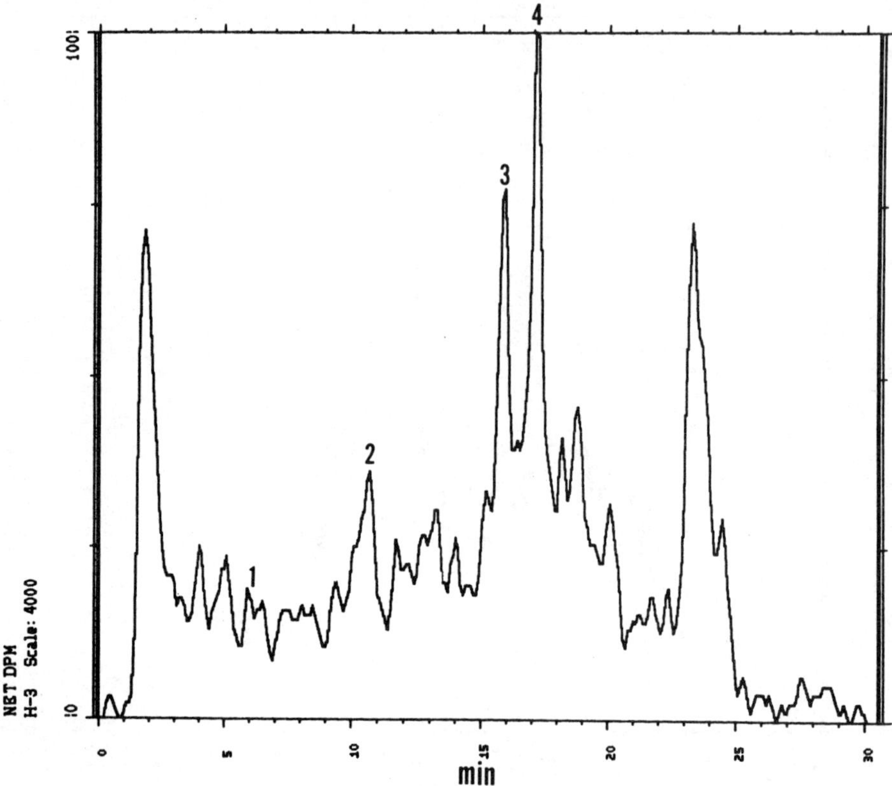

Figure 6b. Radiochromatogram of the MC of the same sample as in Figure 6a. The crosslinks, DHLNL and HLNL (3,4) were better resolved in this figure despite the high level of radioactivity in the peaks. The precursors DHNL and HNL were also present (1,2).

Increasing the organic beyond 36% brought the late-eluting PITC reagent peaks towards the crosslink region and K. Use of a 60:40 acetonitrile:water solution as B did not improve the separation, nor did using a 10% to 51% gradient of the latter as B. Increasing the temperature to 60°C also moved the PITC reagent peaks closer to K and presumably would diminish column lifetime.

Use of the Biocalc program to test different chromatographic conditions such as column number, size, particle size, flow rate and sample molecular weight was helpful in selecting the correct conditions prior to actually performing the chromatography (Stadalius et al., 1985a,b).

Reversed-phase separation of crosslinks and discussion of results

In the past two years, two methods have been published concerning the use of reversed-phase packings for the separation of reducible collagen crosslinks (Reiser and Last, 1983; Smolenski et al., 1984). Reiser and Last utilized orthophthalaldehyde (OPA) as a postcolumn derivatizing reagent and detected crosslinks by fluoresence and/or radioactivity after NaB^3H_4 reduc-

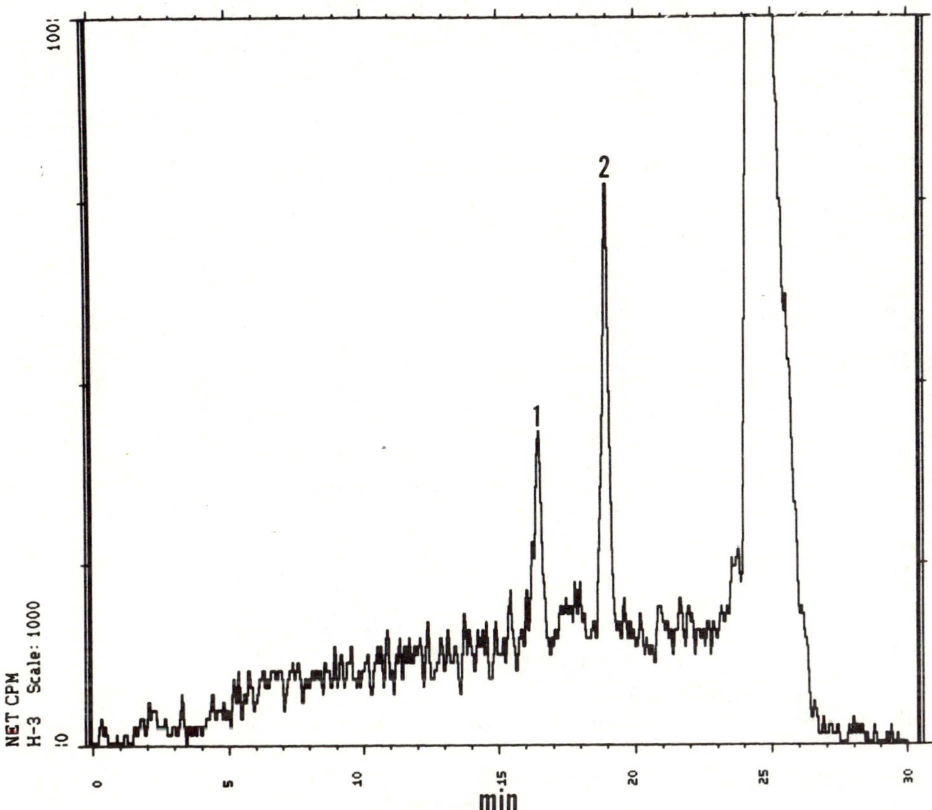

Figure 7a. Radiochromatogram of a 2N NaOH hydrolysate of normal MC from a 3 month-old chick tibiotarsus diaphysis. The sample was processed through a Sep-Pak C18 reversed phase packing prior to derivatization to desalt the sample and remove ^3H_2O. The major crosslink peaks, DHLNL and HLNL (1,2) were present. Note that the prominent artifactual peak that occurs during reequilibration was still present.

tion of the sample (Reiser and Last, 1983). Their system involved a 25 cm x 0.46 cm Ultrasphere ODS column and a gradient of A (0.01 M Na_3PO_4, (pH 2.84), 0.3% sodium dodecyl sulfate and 1.5% n-propanol), and B (35% n-propanol); 0% B for one minute followed by a gradient of 0% to 87% B in 90 minutes (Reiser and Last, 1983). Initial flow rate was 0.7 ml/min up to 25 minutes, then 1.0 ml/min for the remainder of the separation. The temperature of the NaOCl reaction coil was 55°C. DHLNL eluted first at about 62 minutes and HLNL eluted at 65 minutes (Reiser and Last, 1983). Sample masses were in the 200 μg range, but larger samples could most likely be loaded. Although this method was sufficiently sensitive to detect derivatized crosslinks without prior reduction with NaB^3H_4, it involved separate addition of NaOCl to oxidize the secondary amines, proline and hydroxyproline, so that they would react with the OPA. The separation time required more than 100 minutes and did not achieve baseline separation for DHLNL and HLNL.

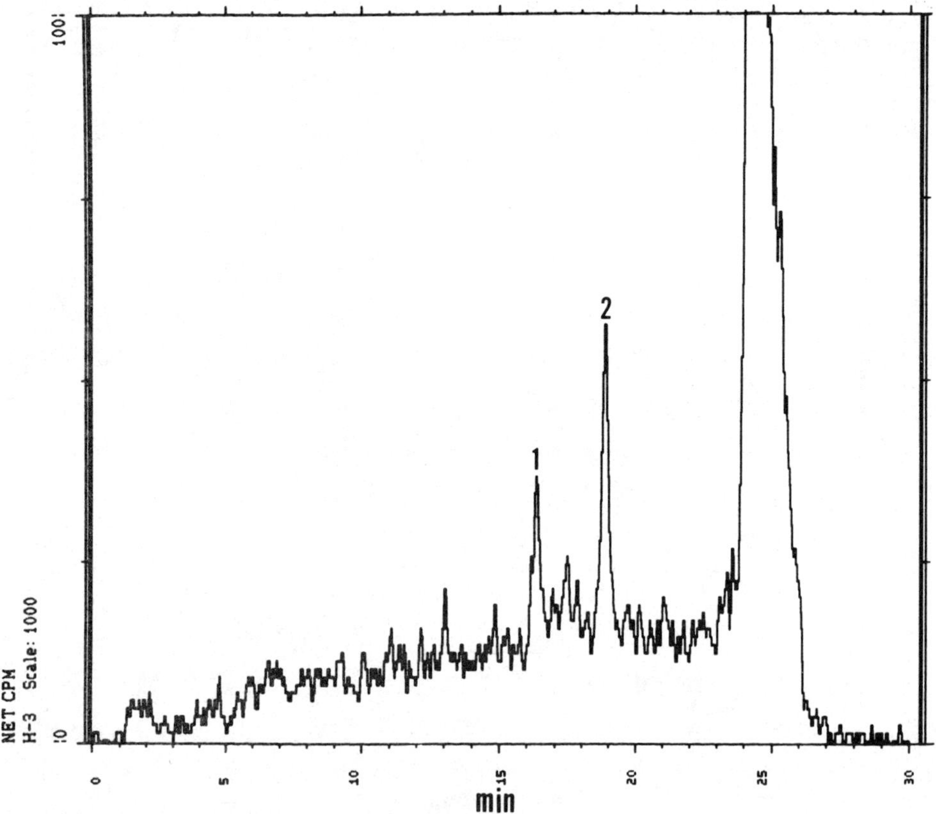

Figure 7b. Radiochromatogram of a 2N NaOH hydrolysate of osteoblastoma MC from a 3 month-old tibiotarsus diaphysis. The sample was processed as in Figure 7a. The crosslink peaks for DHLNL and HLNL were evident.

Another recent method involved separation of the amino acids and crosslinks on a Du Pont, Zorbax amino column at 35°C (25 cm x 0.46 cm NH_2 column, Du Pont) and eluting at 2 ml/min with a complex gradient of A (0.01 M KH_2PO_4, pH 4.3) and B (acetonitrile-water, 500:70 or approximately 88% CH_3CN, 12% H_2O); 80% to 60% B in 65 minutes, where HLNL eluted first at about 40 minutes and DHLNL eluted second at about 44 minutes (Smolenski, 1984). Sample masses were in the range of 1-2 mg. This analysis did achieve baseline separation for the crosslinks, necessitated prior reduction with NaB^3H_4 and required 60 minutes for the complete separation, discounting a one hour reequilibration time at start-up.

For either of the above methods, detection of the radioactive crosslinks labeled in either lysine or with NaB^3H_4 reduction of the Schiff base was performed with scintillation counting of the column effluent. Similar sensitivities were reported with both systems.

The method presented in the current experiments effects elution of the phenylthiocarbamyl (PTC) derivatives of the reducible crosslinks and their

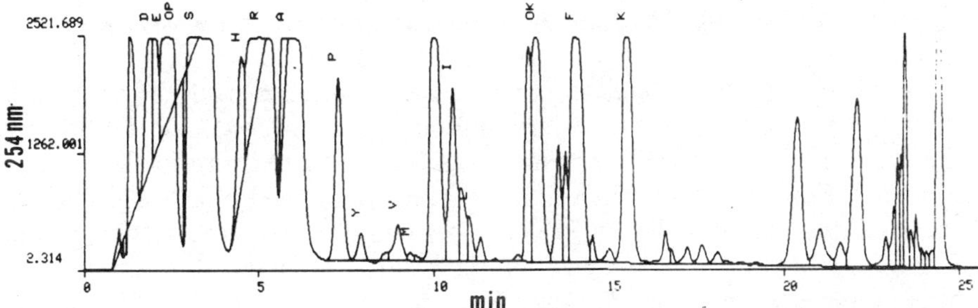

Figure 8a. Absorbance profile at 254 nm of the PTC-amino acids from a separation of a dentin sample used for detection of radioactive precursors and crosslinks. The Du Pont Reliance column yieled good peak shape and peak capacity even though it was grossly overloaded with respect to the PTC-amino acids. The reduced, derivatized, crosslink precursors, DHNL and HNL, eluted before and after I, respectively. The reduced, derivatized crosslink, DHLNL and HLNL, eluted with and after K, respectively.

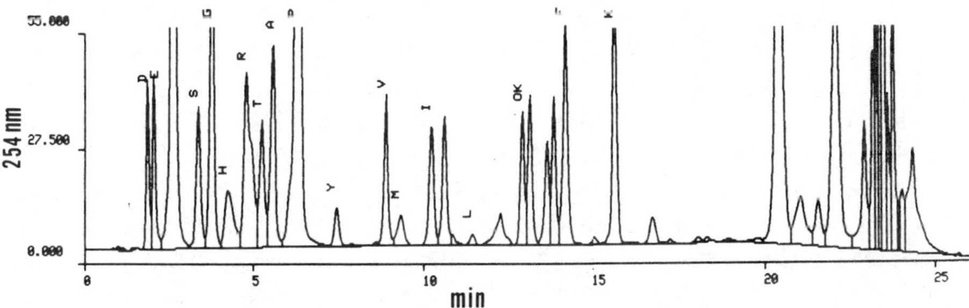

Figure 8b. Analytical separation of the standard PTC-amino acids. OP is clearly resolved from E and S. It was possible to perform both analytical crosslink separations on the same column. However, since in many cases crosslink separation required up to 100 times as much sample mass as did an analytical run, it was deemed best to set up two systems, one for amino acid determinations and a second for the crosslink assays.

precursors in less than 18 minutes. As little as several hundred counts in each crosslink or precursor can be detected using a flow scintillation detector on-line with the column (Fig. 2-6). Sample masses were in the range of μgs to 1 mg, depending on the specific activity of the NaB^3H_4 used to reduce the crosslinks. The crosslinks, DHLNL and HLNL, eluted at 15 and 17 minutes respectively, whereas the precursors, DHNL and HNL, eluted at 6 to 7 minutes and 10 to 11.4 minutes respectively, after mixing with scintillation fluid in a mixing tee and passing through a 500 μl or 2500 μl Teflon coil counting cell in the scintillation counter (Fig. 2-4). Quantitation of radioactivity was updated every six seconds. HLNL and DHLNL eluted at and just after lysine respectively, in a 7% to 36% acetonitrile (B) gradient where mobile phase A is 0.14 M phosphate, (pH 5.7) (see Methods section). Interference by K and a reagent peak that eluted near the crosslink elution position made quantitation by absorbance at 254nm difficult (Fig.8a). One great advantage of this technique is that it requires the least time and materials of any procedure published to date for the detection of collagen crosslinks and precursors.

There are three functional groups that are derivatized by PITC: two alpha amino groups and one secondary amine from the reduced Schiff base (Fig. 1b). The latter nitrogen bears a labile tritium atom from the NaB^3H_4 reduction (Fig. 1b). This tritium on the nitrogen is freely exchangeable with protons in the medium. When PITC reacts with the crosslink at this position, the tritium atom or the exchanged hydrogen atom is displaced, resulting in one tritium atom on the crosslink (Fig. 1b).

The tissues that contain characteristic reducible crosslink profiles such as dentin (DHLNL), rat tail tendon (HLNL) and bone (HLNL and DHLNL) were used to calibrate the column. Samples with a known crosslink content that had been previously chromatographed in a cation exchange system were used as calibration standards for the phenylthiocarbamyl crosslink derivative separations. Human, adult, dentin mineralized collagen gave a typical pattern of one major peak in the precursor region (DHNL) and one major peak in the crosslink region (DHLNL) (Fig. 2). The radioactive peak eluted after a nonradioactive hydroxynorleucine standard (due to the difference in time and the size of the counting coil in the scintillation counter HLNL (Fig. 3b). Reduced, adult, avian flexor tendon had both DHLNL and HLNL crosslinks in the lysine region and a major precursor at the HNL region with a lesser amount of DHNL (Fig. 3a). Reduced bovine periodontal ligament that had been chromatographed in an ion exchange system was also used (Fig. 3c). This sample had the crosslinks DHLNL, HLNL and histidinohydroxymerodesmosine (HHMD) in known amounts. Epiphyseal cartilage gave a more complex pattern but still had a peak at the position of DHLNL as well as DHNL (Fig. 4).

The bone samples contained both DHLNL and HLNL (Fig. 5a,b;6a,b). These samples were divided biochemically into the nonmineralized (NMC) and mineralized (MC) collagen compartments (Banes et al., 1983). Samples from both 3 month-old normal and osteoblastoma birds were prepared. The normal sample had less total crosslinks than did the osteoblastoma sample as had been published previously (after analysis using cation exchange chromatography) (Banes et al., 1978; Table 1). The reducible crosslinks were also unequally distributed in the NMC compared with the MC: there were less crosslinks detected in the MC than in the NMC in both the normal and osteoblastoma samples (Table 1).

One of the metastable, maturation crosslinks in bone, pyridinoline, may be enriched in the NMC (Banes et al., 1983). This crosslink forms from DHLNL and a lysine or hydroxylysine to form lysinopyridinium (LP) or hydroxypyridinium (HP) (Eyre and Oguchi, 1984). If the NMC has an abundance of the pyridinium crosslink, then it follows that the precursor to it, DHLNL, should also be present in abundance in the NMC. Given the present data, it appears that the NMC is enriched in both HLNL and DHLNL compared with the MC. It was demonstrated through the construction of space-filling models that pyridinoline crosslinks draw collagen chains 50 % closer together than does a ketoamine crosslink (Mechanic et al., 1985). This constriction of the collagen chains, due to pyridinium crosslink formation, occurs near the hole zone of collagen, where adjacent molecules meet head-to-tail in quarter-stagger array. We have hypothesized that this constriction at the hole zone may impede mineralization (Mechanic et al., 1985). If this is the case, then an unequal distribution of the reducible crosslinks may also have some impact on the regulation of mineralization.

ACKNOWLEDGMENTS

The authors would like to thank Mr. Fernando DiMeo and Mr. Robert Tatro for technical assistance. Thanks also to Dr. Mitsuo Yamauchi for the sample of reduced bovine periodontal ligament and to Dr. Linda Levin for providing the human dentin sample. This work was supported by NIH grants AM30952, AM30478, AR38121 and the Dental Research Center Core grant DE 02668.

REFERENCES

Banes, A.J., Link, G.W. and Snyder, L.R. (1985). Comparison of reversed-phase columns for the separation of tryptic peptides by gradient elution. Correlation of experimental results and model prediction. *J. Chrom.* 326, 419-431.

Banes, A.J., Yamauchi, M. and Mechanic, G.L. (1983). Nonmineralized and mineralized compartments of bone: I. Quantitative changes in bone collagen crosslinks in virus-induced avian osteopetrosis. *Biochem. Biophys. Res. Commun.* 113, 975-981.

Banes, A.J., Yamauchi, M. and Moody, R.T. (1983). Reversed phase high performance liquid chromatography method for separation of collagen tryptic peptides. *J. Chrom.* 272, 366-372.

Banes, A.J., Bernstein, P.H., Smith, R.E. and Mechanic, G.L. (1978). Collagen biochemistry of osteopetrotic bone: I. Quantitative changes in bone collagen crosslinks in virus-induced avian osteopetrosis. *Biochem. Biophys. Res. Commun.* 113, 975-981.

Banes, A.J. and Smith, R.E. (1977). Biological characterization of avian osteopetrosis. *Infect. Immun.* 16, 876-884.

Eyre, D.R. and Oguchi, H. (1984). Crosslinking in collagen and elastin. *Ann. Rev. Biochem.* 53, 717-748.

Eyre, D.R., Koob, T.J. and Van Ness, K.P. (1984). Quantitation of hydroxypyridinium crosslinks in collagen by high performance liquid chromatography. *Anal. Biochem.* 137, 380-388.

Eyre, D.R. and Oguchi, H. (1980). The hydroxypyridinium crosslinks of skeletal collagens: their measurement, properties and a proposed pathway of formation. *Biochem. Biophys. Res. Commun.* 92, 403-410.

Fallon, A., Lewis, R.V. and Gibson, K.D. (1981). Separation of the major species of interstitial collagen by reverse-phase high performance liquid chromatography. *Anal. Biochem.* 110, 318-322.

Gallup, P.M. and Paz, M.A. (1975). Post-translational protein modifications, with special attention to collagen and elastin. *Physiol. Rev.* 55, 418-487.

Heinrikson, R.L. and Meredith, S.C. (1984). Amino acid analysis by reversed-phase high performance liquid chromatography: precolumn derivatization with phenylisothiocyanate. *Analyt. Biochem.* 136, 65-74.

Housley, T.J. and Tanzer, M.L. (1981). The separation and amino acid analysis of collagen crosslinks on an extended basic ion-exchange column. *Analyt. Biochem.* 114, 310-315.

Jones, B.N., Consalvo, A.P., Leseur, L., Lovato, S., Young, S., Koehn, J. and Gilligan, J. (1986). High performance liquid chromatography of phenylthiohydantoin and phenylthiocarbamyl amino acids. In: *Modern methods in protein chemistry*. (in Press).

Kessler, M.J. (1982). A sensitive radioactivity detector for HPLC. *Amer. Lab.* 14, 53-63.

Kuboki, Y., Tsuzaki, M., Sasaki, S., Liu, C.F. and Mechanic, G.L. (1981). Location of the intermolecular crosslinks in bovine dentin collagen, solubilization with trypsin and isolation of crosslink peptides containing dihydroxylysinonorleucine and pyridinoline. *Biochem. Biophys. Res. Commun.* 102, 114-126.

Link, G.W., Keller, P.L., Stout, R.W. and Banes, A.J. (1985). Effects of solutions used for storage of size exclusion columns on subsequent chromatography of peptides and proteins. *J. Chrom.* 331, 253-264.

Mechanic, G.L. (1974). Collagen crosslinks: direct evidence of a reducible stable form of the Schiff base delta^{c} dehydro-5,5'- dihydroxylysinonorleucine as 5-keto-5'-hydroxylysinonorleucine in bone collagen. *Biochem. Biophys. Res. Commun.* 56, 923-927.

Mechanic, G.L., Banes, A.J., Henmi, M. and Yamauchi, M. (1985). Possible collagen structural control of mineralization. In: *The chemistry and biology of mineralized tissues*, Butler, W.T. (ed.) *EBSCO Media*, Birmingham, pp. 98-102.

Reiser, K.M. and Last, J.A. (1983). Analysis of collagen composition and biosynthesis by HPLC. LC. 1, 498-502.

Smolenski, K.A. (1984). Crosslinks of collagen. In: *CRC handbook of HPLC for the separation of amino acids, peptides and proteins*. 1, pp. 281-289.

Stadalius, M.A., Quarry, M.A. and Snyder, L.R. (1985). Optimization model for the gradient elution separation of peptide mixtures by reversed phase high performance liquid chromatography. Application to methods development. *J. Chrom.* 327,93-113.

Stadalius, M.A., Gold, H.S. and Snyder, L.R. (1985). Optimization model for the gradient elution separation of peptide mixture by reversed phase high performance liquid chromatography. Verification of bandwidth relationship for acetonitrile-water mobile phase. *J. Chrom.* 327,27-45.

Van der Rest, M., Bennett, H.P.J., Soloman, S. and Gloriux, F.H. (1980). Separation of collagen cyanogen bromide-derived peptides by reversed-phase high performance liquid chromatography. *Biochem. J.* 191, 253-256.

Yamauchi, M., Banes, A.J., Kuboki, Y. and Mechanic, G.L. (1981). A comparative study of the distribution of the stable crosslink, pyridinoline in bone collagens from normal, osteoblastoma and vitamin D-deficient chicks. *Biochem. Biophys. Res. Commun.* 102, 59-65.

Progress in HPLC, Vol. 3, pp. 79-91
Parvez et al. (Eds)

Totally automatic combined system comprising an HPLC apparatus and an on-line radiodetector (FLO-ONE) for routine analysis of prostatic enzyme activities

J.M. LE GOFF and P.M. MARTIN
Laboratoire de Cancérologie Expérimentale UA CNRS 1175, Faculté de Médecine Nord, 13326 Marseille Cedex 15, France

SUMMARY

An original system was developed for routine studies of steroid metabolism in the prostate (5α reductase, 17β dehydrogenase). This system was assembled from simple, commercially available components and combines the advantages of highly reproducible HPLC separation and the counting and calculation rapidity of an on-line radiodetector (FLO-ONE). The advantages of this method are: (1) rapid and precise calculation of the conversion rates of an enzymatic reaction without requiring costly double label techniques, (2) limitation of nonspecific radiodecay of the tracers used (elimination of nonspecific controls), (3) considerably reduced consumption of scintillation liquid in the assay. Total automation leads to uninterrupted operation (24 hours a day), speed of analysis at six samples counted and calculated hourly, and less technical assistance. The minimal operating cost of the system, and the advantages it presents in comparison to a conventional procedure of TLC separation with dual labeling and nonspecific phenomena controls are discussed on the basis of the comparative results of 97 assays carried out with the two methods.

INTRODUCTION

The introduction of radiolabeled molecules in biology led to the intensive development of a methodology for routine enzyme assays (1,2). This methodology is based on the conversion of a tracer (substrate) when placed in the presence of an enzyme in standardized conditions. In this type of assay, the analysis of the reaction involves three basic steps:

1. The separation of unmetabolized substrate from the reaction products
2. The recovery and counting of the different radiolabeled products
3. The calculation of the main reaction constants

Each step requires a technological choice regarding the equipment available on the market. Not only the precision, but also the final cost of the assay will depend on this choice.

The conditions of the choice vary, depending on whether the assay in question is on the research laboratory scale or used for routine analyses. On the experimental scale, a satisfactory assay is characterized above all by the precision of measurement it generates. The cost is not a factor in its attainment. In routine assays, on the other hand, reproducibility and final cost of the assay are the two essential parameters that must be optimized. Furthermore, an assay is often introduced for routine use after it has been developed in the research laboratory. It is, therefore, important to select a technique that leads to a flexible adaptation of the initial conditions for a simple and rapid assay. Thus, the technological choice is formulated very precisely when a routine assay is being developed.

The following are required: (1) obtain an optimal combination of technical and economic performances of the different instruments, (2) consider specific requirements for routine utilization, which must be precise, reproducible, rapid, simple and the least costly. This problem of choice emerged when we developed a routine assay for 5α reductase and 17β dehydrogenase activities. We naturally needed to compare the performances of a classical analytical procedure, combining thin layer chromatographic separation (TLC), double labeling and control of nonspecific metabolism with the performances of an analysis system, combining high performance liquid chromatographic separation (HPLC) and instantaneous counting by an on-line radiodetector (FLO-ONE, Radiomatic Instruments, La Queue-Lez-Yvelines, France).

In light of our personal experience, we will show how the combination of HPLC and the on-line radiodetector present unquestionable technical and economic advantages, which make it an excellent tool for studying an enzymatic reaction. We will also show how such a system can be inexpensively adapted for simple and rapid use (automation) with minimal operating costs, justifying its routine use.

MATERIAL AND METHODS

Equipment. The HPLC system included two high pressure (HP) pumps (Altex A 110), two analytical columns 250 mm x 4.5 mm (Regis, USA) packed with fully end-capped ODS spherical silica (5μm), two guard columns (Brownlee Labs, USA) equipped with rechargeable cartridges (RP C18 15 mm x 3.2 mm, 7 μm), an automatic WISP 721 injector (Waters, France), an Altex 420 program controller with its pneumatic interface. UV detection (facultative) was carried out with a Beckman 163 variable wavelength detector. Radiodetection was accomplished with the FLO-ONE (Radiomatic Instruments et Consommables, La Queue-Lez- Yvelines, France).

The components were connected with switching valves (V), consisting of a Rheodyne HP 7000 6-position valve (V6), a Rheodyne LP 5041 4-position valve (V4) and two Rheodyne LP 5302 3-way valves (V3). The valves were equipped with pneumatic actuators, Rheodyne 5701 and Rheodyne 5300. The 3-way valves were mounted with their release springs and the high pressure valve was motorized (Touzart et Matignon, France).

The autosampler was the Waters Intelligent Sample Processor (WISP) 712. An Altex 420 program controller was installed, equipped with the optional pneumatic interface board. The program was written in File O and automatically reinitiated at each injection by an electrical signal from the WISP. An electrical relay with controlled intervals cut out every other re-initiation signal.

Thin layer chromatography was carried out on 20 cm x 20 cm silica gel plates (Merck). A Berthold thin layer scanner was used to locate radioactive zones. Static liquid scintillation counting was performed with a Beckman 8000 scintillation counter using a dual label (^{3}H-^{14}C) sample counting program.

Reagents. Diethyl ether was analytical grade (Prolabo, France). Organic solvents (Merck) were distilled in the laboratory before use. Water was deionized and triple glass distilled. Labeled steroids were obtained from New England Nuclear (Paris): [1,2,6,7-^{3}H] testosterone (NET 370, specific radioactivity 99 Ci/mmol), [4-^{14}C] dihydrotestosterone (NEC 607, specific radioactivity 57.8 mCi/mmol), [6,7-^{3}H] estradiol (NET 013, specific radioactivity 47.4 Ci/mmol) and [4-^{14}C] estrone (NEC 512, specific radioactivity 57 mCi/mmol). The tracers were repurified before use. Cofactors used were NADPH type III (Sigma 6505) or NAD (Boehringer Grade I).

Enzyme assays. Microsomal preparations were from human benign hyperplastic prostate tissue (BPH) (3). Final incubation volume was 1 ml. The reaction mixture for the 5α reductase assay was 20 mM sodium acetate (pH 5.5), NaCl to constant ionic strength of 0.2 I, 0.5 mM NADPH added in 50 μl of 5 mM Tris-HCl (pH 7) 10 seconds before starting the reaction. The incubation buffer for assaying 17β dehydrogenase was 40 mM Tris-HC1 (pH 8), NaCl to constant ionic strength of 0.2I, 1mM NAD.

Incubations contained 10^{6} dpm ^{3}H-testosterone or ^{3}H-estradiol and were prepared to a final substrate concentration of 0.5 μM in the 5α reductase assay and 0.2 μM in the 17β dehydrogenase assay. Reactions were started by transferring 50 μl of microsomal preparation (0.25 mg protein) to pre-heated tubes containing the incubation buffer and were stopped by freezing in liquid nitrogen.

Pellets were extracted twice in ether, the extract was evaporated to dryness in a rotary evaporator, and the extracts were taken up with 20 μl of an ethanolic solution of carriers (0.25 mg/ml testosterone, dihydrotestosterone and 3α androstanediol for the 5α reductase assay, and 0.10 mg/ml of estradiol and estrone for the 17β dehydrogenase assay).

In TLC procedures, ^{14}C-dihydrotestosterone or ^{14}C-estrone were added for recovery calculations. Twenty microliter deposits were applied to the origin. Development was with dichloromethane/diethyl ether (4/1, v/v) for the

5α reductase assay and chloroform/ethyl acetate (8/2, v/v) for the 17β dehydrogenase assay. Radioactive zones detected with the scanner were scraped off the plates, eluted in methanol and counted for 10 minutes in the static scintillation counter. Nonspecific radioactivity was calculated in control incubations in the absence of the enzyme preparation and subtracted from the incubation values to obtain the true conversion rate for each reaction.

In the HPLC procedure, the extract was evaporated and then taken up with 210 μl of the separation solvent: methanol/tetrahydrofuran/water (40/13/47, v/v/v) for 5α reductase and acetonitrile/water (37/63, v/v) for 17β dehydrogenase. One hundred microliter were injected and reinjection was systematic whenever there was less than 1% conversion.

RESULTS

Chromatography set-up. The on-line radiodetector and the components of the HPLC system are depicted in Figure 1. This set-up was identical in both assays (5α reductase and 17β dehydrogenase). The HPLC separation circuit included two identical columns. In each, an HP pump alternatively delivered a mobile separation phase and then a wash, selected by a 3-way valve (V3), through an analytical column. Samples were injected with the automatic injector at fixed intervals and alternating with the two columns. This alternation was assured by the HP 6-position valve (V6). The radiodetector was also connected alternatively to the exit of each column via a low pressure 4-way valve (V4). When one column effluent was being recorded, the exit of the other was connected to the system for recovery of radioactive waste.

Chromatographic conditions. Table 1 summarizes the chromatographic conditions used to assay 5α reductase and 17β dehydrogenase.

Control. The different valves were controlled by the File O program of

Table 1. Chromatographic conditions for assaying the activities of prostatic 5α reductase and 17β dehydrogenase.

Assay	HPLC solvent system		Pump flows (ml/min)		* Counting efficiency (%)	Counting chamber volume (ml)
	Separation	Washing	HPLC	Scintillant		
5α reductase	Methanol 40% Tetrahydrofuran 13% Water 47%	Methanol	1.4	5.6	30.6	2.5
17β dehydrogenase	Acetonitrile 37% Water 63%	Acetonitrile	2	8	29.7	2.5

Counting efficiency (E) was calculated from the total counts (TC) recorded over 10 minutes by the radiodetector when 10^6 dpm of tritiated substrate is present in the chamber with all pumps stopped. The chamber contained a mixture of HPLC effluent and liquid scintillant in the proportions of the separation. (1/4)

E = TC × 10-5 (%)

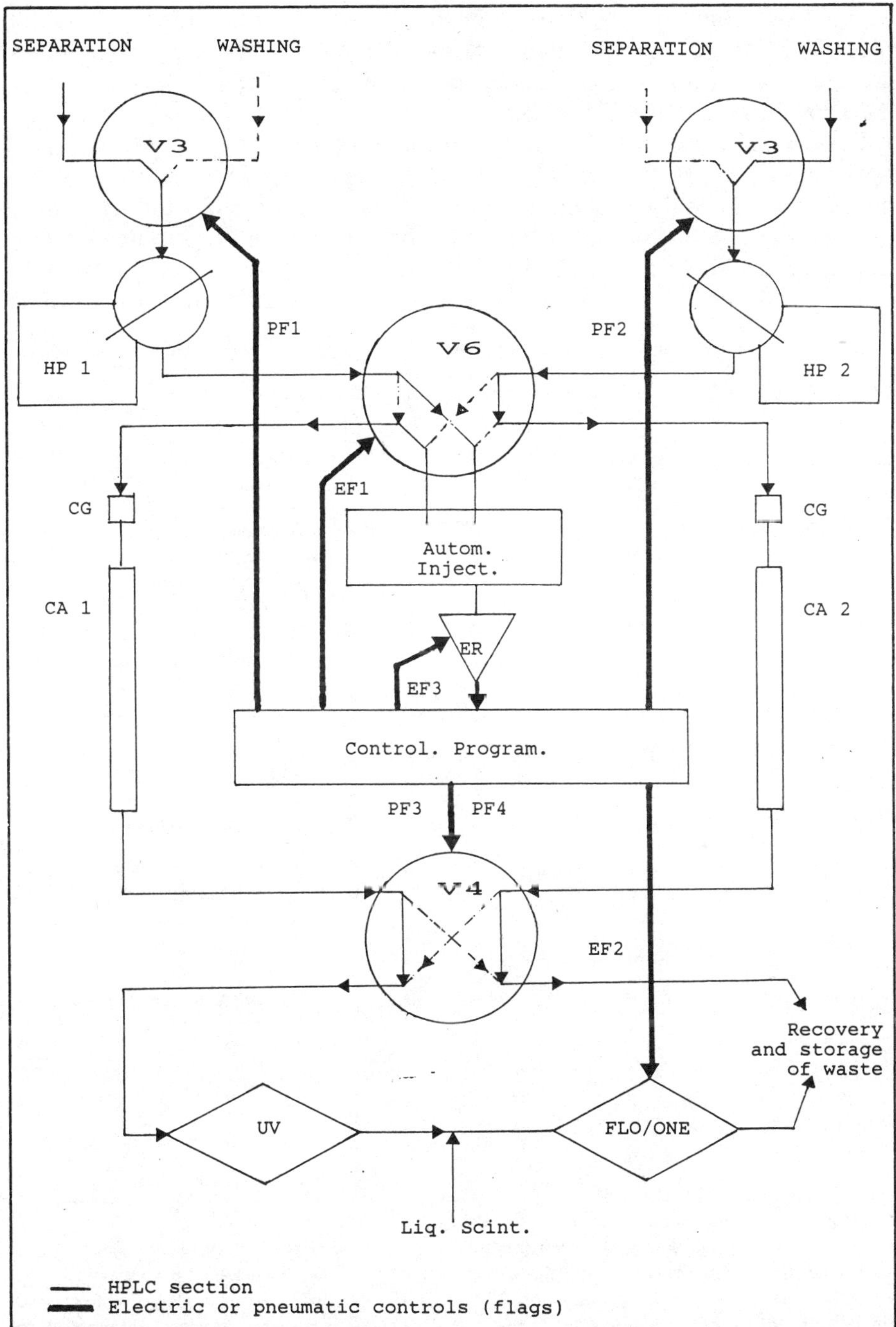

Figure 1. Chromatographic set-up

AC: analytical column; ER: electronic relay; GC: guard column; HP: High pressure pump; UV: UV detector (optional); V: valve;

the program controller, which cyclically activated electrical or pneumatic flags. The chronology and nature of the activations used are shown as an indication in a truth table in the case of our prostatic 5α reductase and 17β dehydrogenase assays (Table 2).

These automatic controls should permit: (1) alternating successive injections on each of the two HPLC separation columns, (2) to detect with the radiodetector only the section of the chromatogram of each sample, which represents the nonmetabolized tritiated substrate and the metabolites formed during the reaction (see discussion), (3) separating simultaneously the next sample on the other column now connected to the system for *recovery and*

Table 2: Truth table for the total automatic control of the HPLC-FLO/ONE system.

Time									
5α	17β	EF1	EF2	EF3	PF1	PF2	PF3	PF4	Comments
0	0								Injection in column 1 starts the program
2	2	1	0	0	0	0	0	0	Injector is connected to column 2
2,8	4	1	0	0	1	0	0	0	Beginning of column 2 wash (variable according to volume between V3 and column 1)
5,10	5,5	1	0	0	1	0	1	0	Column 1 is connected to FLO/ONE
5,30	5,7	1	1	0	1	0	0	0	FLO/ONE stopped
5,40	5,8	1	0	0	1	0	0	0	
5,50	5,9	1	1	0	1	0	0	0	FLO/ONE restarted
5,60	6	1	0	0	1	0	0	0	
5,80		1	0	0	0	0	0	0	End of column 2 wash
6		1	0	1	0	0	0	0	ER relay opened
8	9								Injection in column 2
13	13,5	0	0	1	0	0	0	0	Injector connected to column 1
14	15	0	0	0	0	1	0	0	Beginning of column 1 wash
15,5	15,2	0	0	0	0	1	0	1	Column 2 connected to FLO/ONE
15,7	15,3	0	1	0	0	1	0	0	FLO/ONE stopped
15,8	15,4	0	0	0	0	1	0	0	
15,9	15,5	0	1	0	0	1	0	0	FLO/ONE restarted
16	15,6	0	0	0	0	1	0	0	
17	16	0	0	0	0	0	0	0	End of column 1 wash

Activation (1) or deactivation (0) of the different flags of the program controller.

5α = 5α reductase; 17β = 17β dehydrogenase; EF = electric flag; PF = pneumatic flag; EF1 = command of valve V6 to connect the injector to column 1 (0) or to column 2 (1); EF2 = command to stop (0 –> 1) or start (0 –> 1) the radiodetector; EF3 = command to open electronic relay (ER) to prevent reinitialization of the File O program during injection; PF1 and PF2 = command valves V3 to select the mobile phase for separation (0) or washing (1); PF3 = command valve V4 to connect column 1 to FLO/ONE and column 2 to the system for recovery and storage of radioactive waste (0 –> 1); PF4 = commands valve V4 to connect column 2 to FLO/ONE and column 1 to the system for recovery and storage of radioactive waste (0 –> 1).

storage of radioactive waste (4) synchronizing the operation of two columns to obtain two successive recordings with the radiodetector without losing time, (5) washing each column and reequilibrate after the last product has emerged and before the injection of a new sample.

The program is automatically restarted by an injection signal which is canceled every second time by the opening of an electronic relay, to complete the analysis for two successive injections.

Measurement reliability. 5α reductase activity was assayed in 97 prostatic samples, either with the automatic system or in parallel by the classical procedure of TLC separation, $^{3}H/^{14}C$ double label and control of nonspecific phenomena. There was an excellent correlation (r = 0.96) between the results obtained with the two methods (Fig. 2).

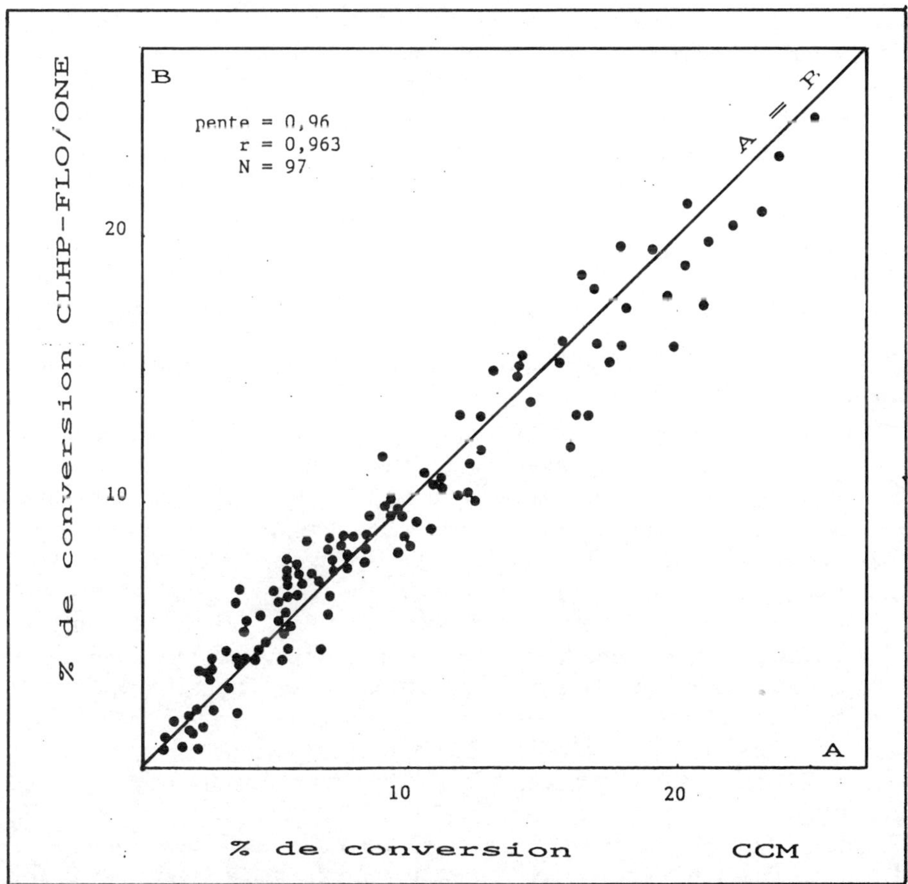

Figure 2. Comparative measurements of conversion rates of testosterone to dihydrotestosterone using two analysis methods.

A = TLC separation. Counting and calculating are performed using dual labeling and nonspecific phenomena controls.

B = HPLC separation. Direct counting and calculation are performed by the FLO/ONE radiodetector.

Measurement precision. Relative error of the measurement was calculated for 37 assays carried out in duplicate with the two methods. No significant difference was found between the two procedures (R.E. TLC = 11.06 ± 9.14%, R.E. HPLC = 12.14 ± 10.64%).

DISCUSSION

When a routine enzyme assay is being developed for measuring the extent of conversion of a radioactive substrate, a choice of technologies must be made for the different steps of the analysis. Usually, the separation technique is chosen from chromatography methods tried and proved in research laboratories which can be easily transferred to an analytical laboratory. Although thin layer (TLC) and paper chromatography (PC) are accessible with no major investment, their technical performances are only average. Liquid chromatography under high pressure, now called high performance liquid chromatography (HPLC), on the other hand, provides superior technical quality and better analytical reproducibility (4,5). These systems are costly, however, and should be introduced into a routine analytical laboratory only if they lead to an appreciable reduction in the cost of the assay and thus, pay their cost rapidly.

The choice of the counting system has long been limited, since only static scintillation counters were available on the market. The use of these counters requires that the reaction products first be recovered in scintillation vials, a time consuming and costly step which increase the cost of technical personnel for the assay, or requires the purchase of specific equipment (fraction collector).

This disadvantage can be overcome by the use of on-line liquid scintillation counters, recently made available on the market. Directly connected to the outlet of the HPLC system, these counters directly and instantaneously recover all the separation products. Some counters, an example being the FLO-ONE (Radiomatic Instruments, La Queue-Lez-Yvelines, France) perform complete counting and recording of the chromatographic profile at the column exit. They also calculate instantaneously the relative percentages of each product contributing to the chromatogram. This technical specificity is particularly valuable when the progression of an enzymatic reaction is studied, as it produces a direct calculation of the conversion rate of the initial substrate in each reaction product. The use of these counters in combination with an HPLC system has been proposed in experimental metabolic research (6).

In light of our personal experience in the field of steroid metabolism, this HPLC/on-line radiodetector combination can be easily introduced for routine use. The HPLC-FLO-ONE combination presents clear technical and economic advantages, which make it the preferred analytical tool for the study of enzymatic reactions.

First, this combination provides considerable precision of measurement at lower cost. It achieves separation without loss and instantaneously recovers the reaction products with a yield close to 100%. In classical procedures with a TLC step, there is a progressive and noncontrollable decrease of reaction products recovery at the various stages of deposit, migration, scraping and elution, preceeding counting. It then becomes necessary to use costly double label techniques or a time consuming recovery and counting of separated metabolites. The use of double label techniques becomes unnecessary by the HPLC-FLO-ONE combination. This system permanently records the entire chromatogram at the column exit. If the metabolites in the incubate are totally extracted and the tracers are completely purified, one obtains a simple, precise and reproducible calculation of the reaction conversion rates. Accordingly, an excellent correlation was found between the results of 97 assays using this procedure or the classical procedure of TLC, double label and control of nonspecific phenomena.

The same combination leads to considerable improvement in the reproducibility of the analysis. This improvement is largely related to the use of an HPLC technique leading to better control of the separation steps and counting. The advantages of this technique over using TLC or PC in the biochemistry of steroids have been described in detailed reviews. (7,8).

This system also limits nonspecific phenomena related to the use of radiolabeled products in our tests. We recently reported difficulties resulting from the use of commercial radioactive molecules with variable radiostability in routine assays (9). The tritiated 17β estradiol used as substrate in the prostatic 17β dehydrogenase assay is difficult to purify and an accelerated radiodecay of the molecule was noted during prolonged contact with a solid phase (silica gel in TLC, ODS in HPLC). Also, complex processes of radiodecay occur during the different steps in TLC (deposit, migration, scraping and elution of products), at ambient temperature and by contact with air. These processes lead to a retention of radioactive products on the plates in the zones where reaction products are expected. This radioactivity is by definition nonspecific, since it is irreproducible because of the extreme experimental variability occuring at different steps in TLC. In the experimental protocols, every tube incubated in the presence of an enzyme must be paired with a tube without an enzyme (nonspecific control) to evalute this spontaneous breakdown, which can lead to erroneous values when calculating the real conversion rate of the reaction. These nonspecific controls add difficulty to the assay, since the number of analyses is doubled, and therefore, their inclusion in a routine assay would double the cost.

It is interesting to mention that the use of an HPLC method limits and controls the spontaneous radiodecay of the tracers. In the case of a highly unstable molecule such as 17β estradiol, the radiodecay of the tritiated metabolites present in the incubate does not exceed 5% at the time of analysis, providing it is carried out within two days of the incubation and that there is a complete (more than 98%), and lasting purification of the tracers (9). Furthermore, the polarities and thus retention times of major radiolysis products are very different from those of the substrate and the reaction products, therefore, the use of nonspecific controls in the assay becomes unnecessary. It is necessary to wash the analytical columns between successive injections to prevent the accumulation and uncontrollable release of radiobreakdown products during analyses. The FLO-ONE on-line radiodetector combined with an HPLC system assures instantaneous collection of metabolites at the column exit and provides optimal reproducibility of the analysis since it permanently limits and controls these nonspecific radiobreakdowns.

Also,this system lowers cost by reducing scintillation liquid consumption, without affecting counting precision. It should be recalled that counting precision decreases as the time of counting decreases. The process of disintegration of a homogeneous population of radioactive sources is described classically by the Poisson law, which approaches the Gaussian law when the disintegration rate is high. The statistical precision of counting thus depends on the mean number of disintegrations recorded by the counter. In addition, the absorption of photons by the liquid phase always limits the number of disintegrations actually recorded by the photomultipliers. This phenomenon is amplified when there are excessive quantities of organic solvents or water mixed with the scintillant. The latter must be used in sufficient proportion to limit the effects of chemical quenching. This also prevents the formation of gels in the counting chamber, which would affect the characteristics of a liquid phase measurement. Thus, in order to obtain high counting precision, counting should be long enough and be carried out in the presence of a sufficient quantity of scintillation liquid.

In the very special case of routine use, an optimum compromise must be found between the choice of counting time and scintillant flow rates to be used. Chromatographic conditions impose a flow rate and a composition of the HPLC effluent, and a minimal scintillant flow rate is necessary to avoid gel formation inside the counting chamber. The scintillant flow rate should naturally be maintained close to the value preventing gel formation, so, that counting sensitivity is not reduced and that cost is not increased. Any increase in flow rate above this value will induce an unnecessary expensive scintillant consumption. Although it leads to a slight increase in the counting rate by decreasing chemical quenching, it also decreases the retention times of radioactive molecules. The counting time of an on-line radiodetector is governed by the total flow rate of the HPLC effluent and the liquid scintillant through the counting chamber, thus, changing scintillant flow rate will very rapidly affect assay sensitivity.

For economic conditions of utilization, the measurement precision of the system can be determined. Counting yield is determined by the composition of HPLC effluent/scintillant mixture. The counting time is defined by the chamber volume and the total flow passing through it. It then becomes possible to calculate the minimal radioactive load that the incubated substrate must have in order for the counting precision of the reaction product to be satisfactory when conversion rates are low (see appendix). In our case, it was necessary to choose tritiated substrate specific activity to be able to incubate 10^6 dpm at the beginning. This generates a counting error less than 5%, even when the minimal conversion value reaches 1%.

Furthermore, recording the total chromatogram makes it possible to study complex enzymatic reactions. It is often very difficult *in vitro* to separate a single enzyme function at the expense of other enzyme complexes which may be present in the same subcellular preparation. A classical example of this is the study of the reduction of testosterone to 5α dihydrotestosterone (DHT) by prostatic 5α reductase. The substrate for this reaction, testosterone, is also the substrate for 17β dehydrogenase, which is the second most active enzyme after 5α reductase in the human prostate. The reaction product, DHT, itself can be converted to 3α androstanediol (3α-diol) by a 3α oxoreductase (10). Although 17β dehydrogenase can be totally inhibited by a judicious choice of pH (in acid medium), it is difficult to prevent a low rate of secondary conversion of DHT to 3α-diol. The 3α oxoreductase is present in highly variable quantities in the samples and so it is of value to use the entire recording of a chromatogram to verify that this secondary conversion remains negligible in the conditions of the assay (3α-diol/DHT less than or equal to 5%). The recording can thus be limited to the outputs of only the metabolized substrate and the reaction product without losing measurement precision. This leads to a considerable reduction of analysis time and liquid scintillant consumption, yielding considerable savings.

HPLC and FLO-ONE thus combine reproducibility and calculation precision for the study of enzyme reaction. Finally, this equipment can be simply adapted to routine use in a rapid and economic manner. A basic set-up of two identical analytical columns mounted in parallel, as described in the above example for the assays of a 5α reductase and 17β dehydrogenase, optimize the operating time of the radiodetector by using a single automatic injector, and washing each column between two successive injections. The mean consumption of scintillation liquid per sample analyzed is reduced to its minimum, since only the section of the chromatogram containing the reaction metabolites is recorded. During the recording time of one sample, the following sample is separated. Analyses are thus rapid (six samples separated, counted and calculated per hour). Simple automation of commercially available components completes the optimization of the routine analysis system with simple uninterrupted operation 24 hours a day, with a minimum of maintenance.

In summary, this automated system is highly reproducible, easy to use, and permanent. It offers a satisfactory analysis precision without the need for double labeling or the control of nonspecific phenomena. In particular, its operating cost is minimal, since it subsequently reduces scintillant consumption, eliminates the operation of product recovery and steps of calculation of reactions. It is thus very well adapted to the routine study of enzyme reactions where radiolabeled substrates are used, and its extremely low operating cost leads to its rapid amortization.

APPENDIX

Calculation of relative counting error in an on-line radiodetector

v = volume of the counting chamber
HPLC = HPLC effluent flow rate
LS = liquid scintillant flow rate
t = counting time
y = counting yield after chemical quenching
TC = total count
CPM = counts per minute
DPM = distintegrations per minute
E = relative counting error (Poisson law)

$E = TC^{+1/2}/TC = TC^{-1/2}$
$TC = CPM.t$
$t = v/(HPLC + LS)$
$CPM = DPM.y$
$E = [(HPLC + LS)/DPM.y.v]^{-1/2}$

A.N.- (17β dehydrogenase)

HPLC = 2 ml/min LS = 8 ml/min Y = 30%

V = 2.5 ml E = 5%

$E \leqslant 5\% \quad \langle = \rangle \quad DPM \geqslant 10/(0.3)(2.5)(0.05)^2 = 5000$ dpm

In order to count with a relative error of less than 5%, 5000 counts must be seen inside the chamber. In the extreme case of minimal conversion (1%), a total of 500,000 dpm must be injected in the column. In order for two successive injections to be made from the same extract, it must contain 10^6 dpm.

The specific activity of the substrate should be 10^6 in the incubation.

REFERENCES

Dixon, M., and Webb, E.C. (1979). (eds). Enzyme Techniques. In: *Enzymes*, Longman Group Ltd. New York. Acad. Press, pp. 21-22.

Ghanadian, R. and Smith, C.B. (1983). Metabolism of steroids in the prostate. In: *The endocrinology of prostate tumors*. Ghanadian (ed). Lancaster, MTP Press, pp. 113-142.

Heftmann, E. and Hunter, I.R. (1979). High pressure liquid chromatography of steroids. *J. Chromatogr*. 165:283-299.

Jayle, M.F. (1961). Application des radioisotopes a la chromatographie des steroides. In: *Analyse des steroides hormonaux*, Masson et Cie (eds). I, Paris, Brodard-Taupin, pp. 205-215.

Kessler, M.J. (1982). High performance liquid chromatography of steroid metabolites in the pregnenolone and progesterone pathways. *Steroids*. 39:21-33.

Le Goff, J.M., Berthois, Y. and Martin, P.M. Radioligand purification prior to routine receptor assays. *J. Steroid Biochem*. (submitted).

Martin, P.M., Le Goff, J.M., Brisset, J.M., Ojasoo, T., Husson, J.M. and Raynaud, J.P. (1987). Uses and limitations of hormone, receptor and enzyme assays in prostate cancer. *Progr. Clin. Biol. Res.* 243 A: 111-140

Robinson, A.L. (1979). HPLC: The new king of analytical chemistry. *Science*. 203:1329-1332.

Rosset, R., Claude, M. and Jardy, A. (1982). La chromatographie en phase liquide. In: *Manuel Pratique du Chromatographie en Phase Liquide*. Masson (ed). *Imprimerie Jouve*, pp. 1-11.

Vestergaard, P. (1973). Liquid column chromatography of hormonal steroids. In: *Modern Methods of Steroid Analysis*. Heftmann (ed). New York, Acad. Press, pp. 2-35.

Progress in HPLC, Vol. 3, pp. 93-127
Parvez et al. (Eds)

HPLC and on-line radioactivity monitoring in studies of vitamin D metabolism

SREEKUMAR PILLAI, ELAINE GEE and DANIEL D. BIKLE
Departments of Medicine and Dermatology, Veterans Administration Medical Center, San Francisco, California USA

SUMMARY

Vitamin D, a steroid hormone critical for regulation of calcium homeostasis, is produced in the skin by the action of ultraviolet light. The most biologically potent form of the vitamin 1,25 dihydroxy vitamin D_3 ($1,25(OH)_2D_3$) is synthesized by subsequent hydroxylations in the liver and kidney. However, a large number of other vitamin D metabolites have been described with various biological activities. Vitamin D and its metabolites constitute the vitamin D endocrine system. One of the major problems in studying the vitamin D endocrine system has been the difficulty in separating and quantitating the many metabolites of vitamin D, but recent advances in high pressure liquid chromatography (HPLC) methodology have made this possible. In this review, we have discussed our experience with HPLC coupled to an on-line radioactivity monitor for studying the metabolism of vitamin D, with particular emphasis on our recent observations of vitamin D metabolism by the skin and kidney.

We have observed that human foreskin keratinocytes *in vitro* metabolize $25(OH)D_3$ to a number of metabolites, including $1,25(OH)_2D_3$. Vitamin D_3 so formed remains mostly in the cell and does not accumulate in the medium under the conditions of these experiments. With time this metabolite is catabolized and more polar metabolites appear in both the cells and the medium. The identity of this metabolite as $1,25(OH)_2D_3$ was confirmed both by its potency in displacing $1,25(OH)_2D_3$ in the chick cytosol receptor assay and by mass spectral analysis. Parathyroid hormone and isobutylmethylxanthine increase the levels of $1,25(OH)_2D_3$ in the cells by increased production and decreased catabolism. Exogenously added $1,25(OH)_2D_3$ at concentrations as low as 10^{-12} M reduces endogenous $1,25(OH)_2D_3$ production and increase its catabolism. These results suggest that $1,25(OH)_2D_3$ may be formed in the epidermis to regulate vitamin D production by the epidermis and to provide an alternative to $1,25(OH)_2D_3$ production by the kidneys.

We then compared the vitamin D metabolism of keratinocytes to that in liver and kidney homogenates. $25(OH)D_3$ and $1,25(OH)_2D_3$ are metabolized to similar, if not identical metabolites by kidney and skin; the liver does not metabolize $25(OH)D_3$ and $1,25(OH)_2D_3$ extensively. Chronic $1,25(OH)_2D_3$ administration causes an increase in *in vitro* conversion of $25(OH)D_3$ to $24,25(OH)_2D_3$ by kidney homogenates while liver metabolism is unchanged. When we compared the metabolism by renal homogenates of $25(OH)D_3$ and $1,25(OH)_2D_3$ using substrates which differed in the location of the tritium label, we found that substrates labeled in C 23, 24 were metabolized more slowly than substrates labeled in C 26,27.

To establish an *in vitro* cellular model for renal vitamin D metabolism, we evaluated several renal cell lines. Three human carcinoma cell lines, two normal human kidney cell lines, and one opossum kidney cell line were evaluated for their ability to metabolize $25(OH)D_3$. All cell lines produced a metabolite coeluting with $24,25(OH)_2D_3$, but none of them produced $1,25(OH)_2D_3$.

Thus, in these studies we have shown the usefulness of HPLC techniques coupled to on-line radioactivity detection in the rapid screening, identification, and quantitation of vitamin D metabolites produced by various cells. The rapidity, sensitivity and high resolution of these techniques make them ideal for such metabolic studies.

INTRODUCTION

The ability of a fat soluble vitamin to cure the disease rickets was recognized at the beginning of this century (Mellanby, 1918). In 1937 the natural form of vitamin D (vitamin D_3) was isolated and identified from irradiated mixtures of 7-dehydrocholesterol (Windaus and Boch, 1937). More recently, it is appreciated that vitamin D_3 is not only a vitamin but a steroid hormone produced in the skin. Vitamin D_3 is produced from 7-dehydrocholesterol in the skin by a two step process; under the influence of UV irradiation, 7-dehydrocholesterol is converted to pre-vitamin D_3, which is then converted to vitamin D_3 by thermal isomerization. (Holick et al, 1979; 1980; Mac Laughlin et al, 1982). Vitamin D_3 is subsequently released from the skin to the circulation and accumulates in the liver where it undergoes 25-hydroxylation to form $25(OH)D_3$ (De Luca, 1983,). Further metabolism of $25(OH)D_3$ occurs primarily in the kidney where a variety of metabolites are produced (Fraser and Kodicek, 1970), the most important of which are 1,25 dihydroxyvitamin D_3 ($1,25(OH)_2D_3$) and 24,25 dihydroxyvitamin D_3 ($24,25(OH)_2D_3$) (Kulkowski et al, 1979). Recent studies suggest that skin also participates in the further metabolism and function of vitamin D (Nemanic et al, 1985; Bikle et al, 1986a, 1986b). Skin cells possess receptors for $1,25(OH)_2D_3$ (Clemens et al, 1983; Stumpf et al, 1979) and respond to $1,25(OH)_2D_3$ by increasing their state of differentiation (Smith et al, 1986; Hosomi et al, 1983). Furthermore, like kidney cells, skin cells can produce $1,25(OH)_2D_3$ and $24,25(OH)_2D_3$ from $25(OH)D_3$ (Bikle et al, 1986).

One of the major problems in studying vitamin D metabolism has been the difficulty in separating and quantitating the many metabolites of vitamin D. A number of methods have been used toward this end. Most of these methods involved complex pretreatment steps, extractions with various solvents, and multiple types of chromatography. Part of this problem was solved by the introduction of HPLC, which permitted the separation of different metabolites by a single column in a short period of time. Quantitation remained a problem because of the low levels of these metabolites under physiologic conditions. Synthesis of high specific activity radiolabeled metabolites of vitamin D made it possible to study the interaction of vitamin D with cells and organelles under physiologic conditions, but these studies were still hindered by the requirement for fraction collection and counting of individual samples. Therefore, we employed an on-line radioactivity detector in these studies. In this review, we discuss our experience with the use of radioactive flow detectors coupled to HPLC in the study of ^{3}H-vitamin D metabolism by various cellular preparations.

METHODOLOGY

1. Review of older methodology

Until the late 1950's, the only reliable method for the determination of vitamin D in biological fluids was the rat line test (Mc Collum et al, 1922), which depended on curative effects of vitamin D and its biologically active metabolites in rats maintained on a rachitogenic diet. These assays are time-consuming, insensitive, not totally specific, and difficult to quantify precisely. A chemical method was later developed in which vitamin D was reacted with antimony trichloride and the colored reaction product was quantitated by spectroscopy (De Luca and Blunt, 1971). Subsequently, other calorimetric or spectrophotometric methods were developed, which involved reactions with tetrachloroethane (De Luca and Blunt, 1971). Purification of vitamin D by thin layer chromatography (Bollinger, 1971), gas chromatography, (Sheppard and Hubbard, 1971), column chromatography (Strong, 1976), successive column and thin layer chromatography or paper chromatography (Bollinger, 1971) were often used prior to the final calorimetric or spectrophotometric determination to improve specificity. These techiniques have been extensively reviewed and summarized (Hashmi, 1972). Such chemical methods are tedious and insensitive. The advent of protein binding assays and radioimmunoassays has made quantitation of vitamin D and its metabolites a more rapid and sensitive procedure. But these assays are not specific for individual vitamin D metabolites, so that separation of the metabolites, generally by HPLC, must still be used to purify these metabolites prior to assay.

2. Review of current methods

Several excellent reviews are available which describe the different HPLC methods currently used for the isolation and quantitation of vitamin D and its hydroxylated derivatives in biological samples (Fastre and Van Haelen, 1980; Horst, 1983). Depending on the biological sample, several techniques are used for dispersion, solubilization, extraction and prepurification of vitamin D metabolites before HPLC. For example, for estimations of vitamin D content of fish oils, saponification procedures are used to liberate vitamin D from its conjugates (Egaas and Lambertsen, 1979). The vitamin D is extracted from the saponified mixture by organic solvents such as n-pentane, m-hexane, chloroform, diethylether or benzene. In some other cases, as for extraction of vitamin D from serum samples, direct organic solvent extraction without saponification is used (Horst, 1983). The extraction procedure for tissue that we use is as described by Bligh and Dyer (1959) in which the biological sample, chloroform (containing 0.05% butylated hydroxy toluene [BHT], as an antioxidant), and methanol are combined in the ratio 1:1:2 (v/v) to form a monophasic system, which provides maximal contact between the samples. Addition of one part chloroform and 0.8 part water separates the mixture into two phases. The lower chloroform layer contains vitamin D and its biologically active metabolites; the upper aqueous methanol layer contains more polar metabolites for which no biological activity has been identified.

Adsorption HPLC chromatography on microparticulate silica gel is often used for the separation of vitamin D metabolites. Selectivity is good with 10 μm particles, but efficiency is increased with 5 μm particles. Different mobile phases have been used for adsorption HPLC depending on the application. Generally, dichloromethane or hexane in combination with isopropanol is used for straight-phase HPLC, and acetonitrile or methanol with water is used for reverse-phase HPLC. Several commercially available columns such as Partisil (Whatman), Zorbax Sil (Du Pont) or μ-Porasil (Waters) can be used effectively for the separation of vitamin D metabolites using straight-phase solvent systems, whereas $C_{18}\mu$-Bondapak (Waters) and Zorbax ODS (Du Pont) columns can be used for reverse-phase HPLC.

Due to the high selectivity of HPLC, little or no prepurification is required for samples assayed by radioactivity monitoring, although some prepurification steps are required when the subsequent analysis employs competitive protein binding assays. HPLC is often used for the purification of vitamin D and its hydroxy metabolites before their structural identification by mass spectrometry or other physical or chemical methods. In contrast with other methods such as gas chromatography, HPLC does not require derivatization. Thus, HPLC is the best nondestructive method for rapid purification and quantitation of vitamin D and its metabolites.

3. *Detection techniques and automation*

Detection of vitamin D and its metabolites can be performed by UV spectroscopy using an absorption wavelength of 254 or 265 nm. The use of electrochemical detectors may also be feasible because vitamin D has been assayed *in vitro* by oscillopolarography (Menicagli et al, 1976) and voltametry (Atuma, 1975). A peak height and/or area is used to quantitate the amount of vitamin D in a sample by comparison to a corresponding authentic standard. These methods are generally not sensitive enough to detect the levels of vitamin D and its metabolites found in most biological samples. The most sensitive techniques for detection of vitamin D and its metabolites require radioactively labeled derivatives. Vitamin D metabolites labeled with tritium are commercially available, permitting their widespread use.

Before the introduction of HPLC, the methods used for vitamin D analysis were not amenable to automation. Since automatic injectors, programmed solvent delivery systems and automatic data integrators are available for HPLC integration, it is now possible to automate vitamin D analysis. The different steps of HPLC including the formation of the solvent gradient, the pre and post run column equilibration, sample injections and data analysis can all be programmed. Programmable radioactivity flow detectors that can be included in such systems have been developed.

4. On-line radioactivity detectors in HPLC

Detection and measurement of radioactivity in the eluent from a liquid chromatography column generally requires the collection and counting of successive fractions. This process is time-consuming since in HPLC large numbers of small fractions must be collected and counted in order to maintain resolution. Therefore, radioactivity determinations can extend the duration of analysis from minutes to hours or days, and as a consequence, much of the advantage of HPLC is lost. For this reason, continuous flow monitoring devices for radioactivity are advantageous. To allow us to measure tritium (a low energy beta emitter), we selected a system in which the eluent from HPLC is mixed with an appropriate liquid scintillation cocktail and directed to a flow cell placed between two photomultiplier tubes. The detectors respond to the light emitted from the scintillant, which is proportionate to the amount of radioactivity flowing through the flow cell. Other systems utilize a solid phase method, but to date, the efficiency of such systems for tritium is low. The detector integrates the radioactivity of a large number of discrete independent measurements of samples equal in size to the volume of the flow cell averaged over a period of time (six seconds in this case). When maximum resolution is required, as in analysis of complex mixtures, flow cell size can be reduced so that band spreading is avoided. In situations requiring maximum sensitivity, flow cell volume can be increased and flow rate lowered. The radioactivity monitor can be programmed to determine radioactivity at six second intervals, plot the peaks, calculate the radioactivity within each peak and determine the fraction of the radioactivity in each peak relative to the total amount of radioactivity in the sample.

5. Instrument set-up

In all experiments we describe here, HPLC is carried out with a Waters Associates HPLC Model LC-204 chromatograph fitted with two Model 6000A pumps and a Model 440 UV-fixed wavelength (254 nm) detector. The detector response is recorded with a Houston Instruments dual pen recorder. The samples and standards are introduced onto the HPLC column automatically with a 710A Waters "Intelligent" autosampler (WISP 710A). The procedure involves transferring the samples into a limited volume insert

(Waters Part No. 72704) with a final volume of 150 μl. Introduction of the samples to the column in this volume results in minimal peak spreading with negligible sample losses (dead volume of insert is 6 μl). When collection of individual fractions is required, a programmable fraction collector is used. The individual fractions were collected either by time or by volume. The fraction collector is connected to the WISP whereby the individual fractions are collected only while the sample is running through the HPLC. During column re-equilibration, the fraction collector is turned off automatically.

For the quantitation of the radioactivity in the samples, a FLO-ONE Model HP radioactive flow detector, available from Radiomatic Instruments and Chemical Co., Tampa, Florida, is used. The radioactivity detector is connected in series with the UV detector. The eluent from the HPLC column is mixed in a special chamber with a high efficiency nongelling scintillation fluid using a valveless, piston pump. The mixture is then pumped through a transparent coil positioned between two photomultiplier tubes where the radioactive events are measured. A strip chart recorder is attached to the detector and digital rate meter, so that numeric data are continually recorded. The digital information recorded includes retention time, activity in each peak (as cpm or dpm), total activity and percent distribution of activity in each peak. The radioactivity detector is also connected to a dual pen chart recorder which simultaneously monitors the UV profile as well as the radioactivity profile of the HPLC effluent. FLO-SCINT (Radiomatic) liquid scintillation solutions are generally used but other solutions can be substituted.

The solvents used for HPLC are UV spectral grade. Prior to use, solvent mixtures are degassed five to ten minutes to prevent air from entering the HPLC pumps. The columns used for HPLC are purchased from Du Pont (Wilmington, Delaware) or Waters (Milford, MA). The Zorbax-Sil column (Du Pont) is equivalent to the μ-Porasil column (Waters) and Zorbax ODS column (Du Pont) is equivalent to the μ-Bondapak C-18 column (Waters). The Zorbax-Sil and μ-Porasil columns are used for straight-phase HPLC; the Zorbax ODS and μ-Bondapak C-18 columns are used for reverse-phase HPLC.

6. Determining the efficiency and sensitivity of the radioactivity monitor

Liquid scintillation spectroscopy used in the setting of HPLC monitoring requires special counting solutions that minimize chemiluminescence and quenching with the solvents employed for HPLC. FLO-SCINT I is recommended for use with organic solvents such as those used with straight-phase HPLC of vitamin D metabolites. A ratio of 1:1 to 1:2 of effluent to FLO-SCINT is recommended by the manufacturer. We have carried out initial experiments to determine the ratio of effluent to FLO-SCINT that would maximize counting efficiency but minimize scintillation fluid (and cost).

Oxi-test HO standards were purchased from Radiomatic Instruments and Chemical Co. Each capsule contained 0.05 μCi glycerol tri-palmitate-9,20 ^{3}H per capsule. The contents of the capsule were solubilized in hex/IPA (90:10) and mixed with FLO-SCINT I in the following proportions:

	FLO-SCINT I	*HEX: IPA 90:10*
A	1	0
B	3	1
C	2	1
D	1	1
E	1	2

0.01 μCi of ^{3}H standard was mixed with 20 ml of each of the above solvents. Each of the solutions was injected into the counting chamber of the FLO-ONE detector and counted for one minute. This was repeated 10 times for each solution. The average cpm was used to calculate efficiency. The efficiency of counting was the quotient of the cpm measured and the dpm in the 500 μl volume of the flow cell.

$$\% \text{ Efficiency} = \frac{\text{cpm obtained} \times 100}{\text{dpm in } 500\ \mu\text{l cell vol}}$$

FLO-SCINT alone gave an efficiency of 23.7%, whereas a ratio of FLO-SCINT to hex/IPA of 1:2 gave a counting efficiency of 8.9%. A 1:1 mixture showed 15% efficiency and 2:1 mixture 19%. Increasing the ratio of FLO-SCINT to hex/IPA to 3:1 increased counting efficiency to 22%. This increase in efficiency from 19% to 22% was not felt to be worth the cost of the additional scintillant, so in all studies using radioactivity detection we used a ratio of FLO-SCINT to eluent of 2:1. Subsequent tests of efficiency indicated that it remained between 18% - 21%. It should be noted that dichloromethane, another commonly used solvent for HPLC, markedly quenches ^{3}H radioactivity and cannot be used for on-line radioactivity detection.

In order to determine the sensitivity of the radioactivity monitor, we injected different known amounts of radioactive [^{3}H] 25(OH)D_3 and [^{3}H] 1,25$(OH)_2D_3$ into the HPLC system and monitored their elution by the detector. HPLC was carried out using a non-linear, concave gradient from 97:3 to 90:10 hexane:isopropanol at 2 ml/min flow. FLO-SCINT I was the scintillation fluid, and it was mixed with the effluent in a 2:1 (FLO-SCINT:effluent) ratio. The detector was programmed for an efficiency of 19% and a background count rate of 90 dpm (which was previously determined using a blank injection). This background is independent of the flow rate (when the flow is stopped, the same count rate is obtained). Therefore, the contribution of the background is considered as the static count rate. The background count rate is automatically subtracted from the dynamic count rate which is a function of the size of the flow cell, the rate of flow, as well as the radioactivity in the sample. The FLO-ONE Model HP detector calculates the cpm values using the equation:

$$\text{NET cpm} = \frac{\text{counts at 6 second interval x flow rate}}{\text{cell volume}} - \text{background cpm}$$

Failure to subtract the proper background cpm alters the calculated count rate. Therefore, it is essential to determine a proper background value since an improper setting can result in erroneous values especially for peaks with low amounts of radioactivity. As seen in Table 1, the radioactive flow detector was accurate for metabolite peaks that contained at least 500 dpm.

APPLICATIONS

1. Vitamin D metabolism in cultured human foreskin keratinocytes

a. Introduction. Vitamin D_3 is produced primarily, if not exclusively in the kidney under normal physiologic circumstances (Fraser et al, 1970, Reeve et al, 1983). However, some studies suggest that $1,25(OH)_2D_3$ may also be produced by human bone cells in culture (Howard et al, 1981), in melanomas (Frankel et al, 1983), in sarcoid tissue (Mason et al, 1984), and in placenta (Gray et al, 1979). Furthermore, anephric humans (Lambert et al, 1982) and anephric pigs (Littledike et al, 1982) appear to have circulating levels of $1,25(OH)_2D_3$, which are most readily detected after vitamin D_3 or $25(OH)D_3$ administration. Therefore, the possibility that other tissues produce $1,25(OH)_2D_3$ when production by the kidney is reduced needed further consideration. We tested the possibility that cells from the epidermis produce $1,25(OH)_2D_3$ - a mechanism that could provide a means to regulate vitamin D_3 production in the epidermis as well as an alternative source of $1,25(OH)_2D_3$ production in patients with renal failure.

b. Methods. Human newborn foreskins were obtained at circumcision. Keratinocytes were isolated by treatment with trypsin and collagenase and plated onto a monolayer of mitomycin C-treated feeder 3T3 cells in Dulbecco's modified Eagles medium containing 20% fetal calf serum, penicillin, streptomycin, and amphotericin B, according to the method of Rheinwald and Green (1975). Immediately prior to assay, feeder cells were removed from the cultures with 0.1% EDTA. Second or third passage cells were studied after they had achieved confluence. All experiments were done on keratinocyte cultures within one or two days of reaching confluence. At this stage, a few 3T3 cells (<5%) remained attached to the dish under the overlying, mul-

Table 1. Sensitivity of FLO-ONE detector

Injection Vol.	Amount Injected (dpm)		HLPC Peaks Amount Detected (dpm)	
(μl)	$25(OH)D_3$	$1,25(OH)_2D_3$	$25(OH)D_3$	$1,25(OH)_2D_3$
100	1650	1650	1555	1817
66	1100	1100	1155	1077
30	500	500	590	309
20	330	330	116	44
10	170	170	17	0

tilayered islands of keratinocytes. The relative numbers of 3T3 cells and keratinocytes were ascertained by fluorescence microsopy of cultures after treatment with 1 μg/ml of acridine orange (pH 7.0). Whereas acridine orange labeled the cytosol and nuclei of 3T3 cells orange-to-red, keratinocytes were easily distinguished by their green-staining cytosol. Moreover, ultrastructural studies on these cultures revealed that virtually all of the cells in confluent cultures were keratinocytes with desmosomes and bundles of intermediate filaments.

After achieving confluence, the protein and DNA content of the culture dishes remained stable for at least nine days. The mean intra-assay coefficients of variation for protein and DNA content from three experiments, each with six replicate dishes, were 9.9% and 7.3%, respectively. The inter-assay coefficients of variation for protein and DNA content for these three experiments were 4.7% and 3.6%, respectively.

Serum free medium replaced the original medium 40 hours prior to the assay of $25(OH)D_3$ metabolism. In general, 0.1 μCi 25-hydroxy [26,27-methyl ^3H] vitamin D_3 (148-153 Ci/mmol, Amersham Corp., Arlington Heights, IL) was added to the culture dish (3 cm diameter, 1 ml medium) in 10 μl ethanol. Following the desired period of incubation at 37°C under 1 atm 5% CO_2 in air, the reaction was stopped with 1 ml methanol. For most experiments, cells and medium were extracted together by the method of Bligh and Dyer (1959). The aqueous extract was counted for radioactivity without further processing. The chloroform extract was chromatographed using a Waters HPLC system equipped with a Radiomatic FLO-ONE Model HP radioactivity detector, a Du Pont Zorbax Sil column (4.6 mm x 25 cm), and a Waters Systems Controller to deliver a nonlinear (program 9) gradient from 97:3 to 90:10 hexane:isopropanol at 2 ml/min. The radioactivity monitor determined radioactivity at six-second intervals, plotted the peaks, summed the radioactivity within each peak, and expressed each sum as a fraction of the total radioactivity in each sample. The results were converted to fmol metabolite/well, based on the specific activity and the amount of $[^3H]$-$25(OH)D_3$ orginally added to the incubate. This calculation assumes that each metabolite formed has the same specific activity as the substrate. This assumption would not be valid if tritium loss occurred (as would happen for the production of 25,26-dihydroxyvitamin D_3 using the 26,27-labeled $25(OH)D_3$ substrate) during the production of the metabolite. Accordingly, for metabolites whose structure is not known, the results must be considered relative and not absolute. Prior to chromatographic analysis of each group of samples, the column was calibrated with the standards $25(OH)D_3$, $24(R),25(OH)_2D_3$, $1,25(OH)_2D_3$, and for some experiments with $25(R),26(OH)_2D_3$, $1,24(R),25(OH)_3D_3$, and $1,25(R),26(OH)_3D_3$. The details of each experiment are in the figure legends. In these experiments, recovery of radioactivity was 80 ± 5% in the combined aqueous extract and chromatogram of the chloroform extract.

The presumptive $1,25(OH)_2D_3$ peak was identified by cochromatography with chemically synthesized $1,25(OH)_2D_3$ in four different HPLC systems

(7.8 mm × 30 cm μ Porasil column eluted with 9:1 hexane:isopropanol; 46 mm × 25 cm Zorbax Sil column eluted with 97:3 to 90:10 hexane: isopropanol gradient or with 96:4 dichloromethane: isopropanol; and a 3.9 mm × 30 cm C18 μ Bondapak column eluted with 75:25 methanol:water), by its equipotency to chemically synthesized $1,25(OH)_2D_3$ in displacing [^3H] $1,25(OH)_2D_3$ from the chick intestinal cytosol receptor, and by mass spectroscopy. These results are reported in a subsequent section (see below). The presumptive $24,25(OH)_2D_3$ and $1,24,25(OH)_2D_3$ peaks have been identified by cochromatography with chemically synthesized $24,25(OH)_2D_3$ and 1,24(R), $25(OH)_3D_3$ on three HPLC systems (7.8 mm x 30 cm μ Porasil column eluted with 9:1 hexane:isopropanol; 4.6 mm x 25 cm Zorbax Sil column eluted with 97:3 to 90:10 hexane:isopropanol gradient; and a 3.9 mm × 30 cm C18 μ Bondapak column eluted with 75:25 methanol:water) and complete (>95%) loss of radioactivity (the metabolites were labeled in the 26,27 position) following periodate cleavage which cleaves bis glycol linkages such as C24-C25.

c. Results. The addition of [^3H]$25(OH)D_3$ to human keratinocytes resulted in rapid uptake (or binding) of radioactivity by the cells that was maximal by 1 hour of incubation (Fig. 1). After 1 hour, the radioactivity disappeared from

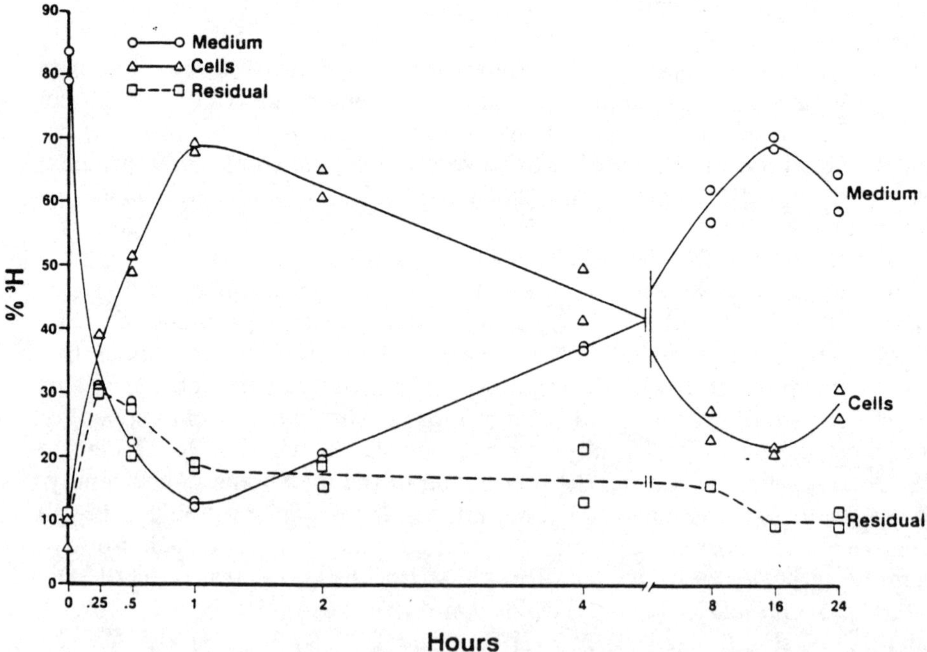

Figure 1. The distribution of radiolabel between cells and the medium as a function of time after the administration of [^3H] $25(OH)D_3$ to the cultured keratinocytes. Duplicate dishes were analyzed at each time point and individual data points are shown. The residual radioactivity is that which was extracted by methanol rinses of the dishes after the medium and cells were removed.

the cells and reappeared in the medium. As seen in Figure 2, most of the radioactivity was in the form of 25(OH)D_3 in the first hour. However, very little 25(OH)D_3 remained after 24 hours. The principal metabolites formed from 25(OH)D_3 after 4 hours of incubation are depicted in the chromatogram in Figure 3. In this figure, peaks with the elution positions of the 25(OH)D_3, 24,25$(OH)_2D_3$, and 1,25$(OH)_2D_3$ standards are so labeled. Two minor metabolites eluted earlier than the 25(OH)D_3, and another metabolite (minor in this chromatogram) eluted in the position of 24,25$(OH)_2D_3$ at 13 minutes. Other metabolites eluted at approximately 22, 23, 24, 27, and 28 minutes in this chromatographic system. Only the metabolite eluting at 23 minutes has been completely identified, and it is 1,25$(OH)_2D_3$ (see below). The presumptive 24,25$(OH)_2D_3$ peak is periodate sensitive, but its structural identity has not been confirmed by mass spectrometry. The standards 25(R),26$(OH)_2D_3$, 1,24(R),25$(OH)_3D_3$ and 1,25(R),26$(OH)_3D_3$ elute at 21, 31, and 35 minutes respectively. Peaks eluting at these times were minor or nonexistent in this

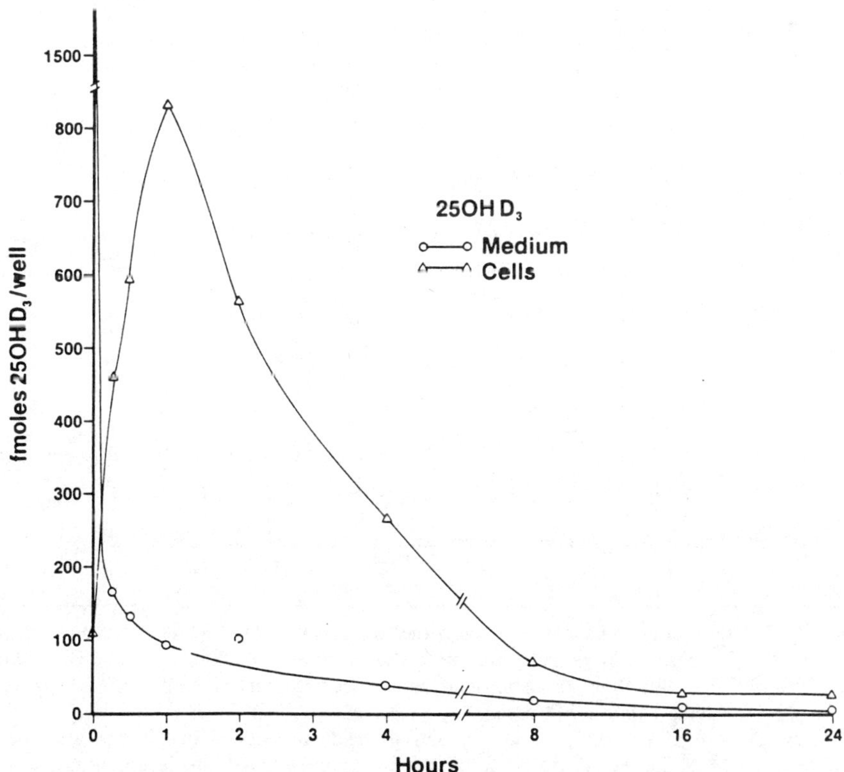

Figure 2. The recovery of 25(OH)D_3 from cells and medium as a function of time following the addition of 1600 fmols of [^{3}H] 25(OH)D_3 to cultured keratinocytes (as in Fig. 1). The medium was removed, the cells were rinsed once, then scraped off into fresh medium. Both cells and medium were extracted separately and analyzed for 25(OH)D_3 by high performance liquid chromatography. The mean of determinations from duplicate dishes is shown.

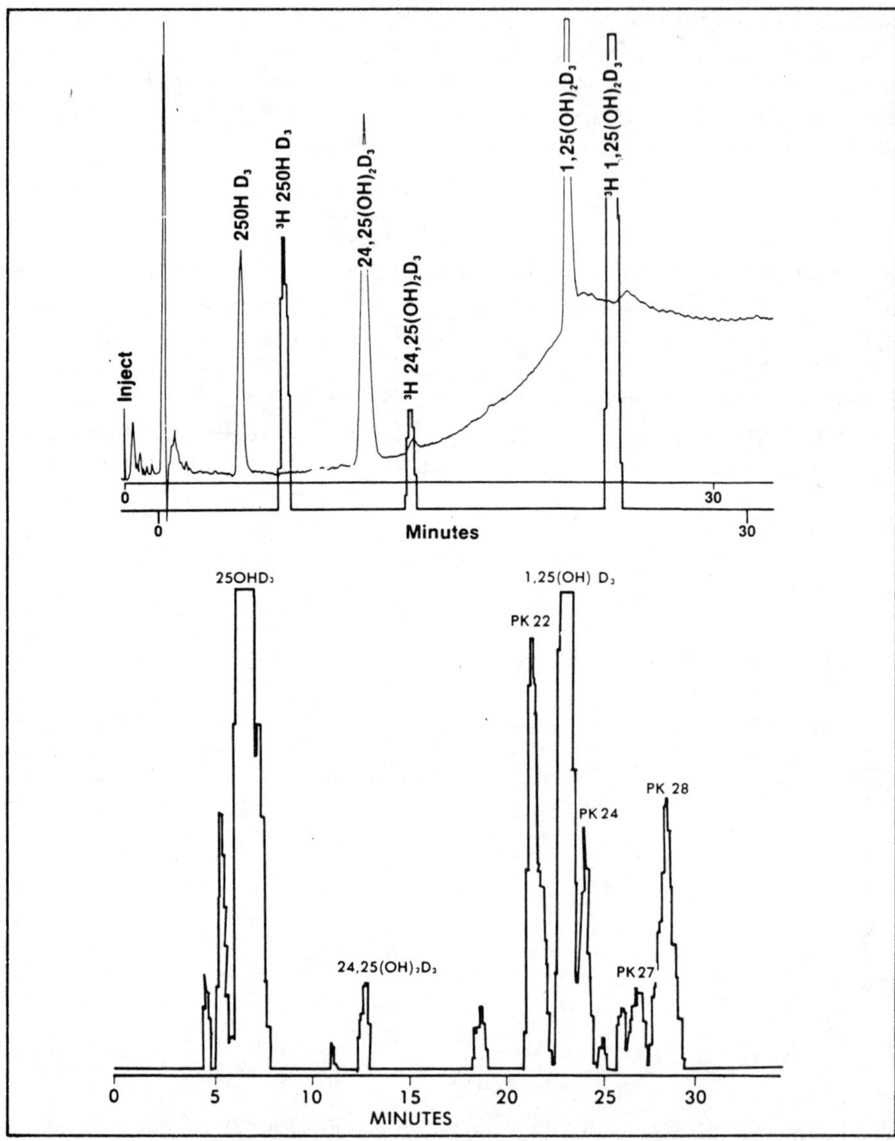

Figure 3. Representative chromatogram of the metabolites of $25(OH)D_3$ produced by keratinocytes. The chromatographic system used was a 4.6 mm × 25 cm, Zorbax Sil column eluted with a 97:3 to 90:10 hexane-isopropanol nonlinear gradient at 2 ml/min with effluent monitored by UV absorbance at 254 nm and by a Radiomatic FLO-ONE radioactivity monitor (dark line). The offset between UV and radioactivity peaks reflects the distance between pens on the recorder (note the different "0" starting points), plus a small displacement due to the volume of tubing (1ml) connecting the monitors. In the top panel, the elution patterns of radioactive and nonradioactive standards $25(OH)D_3$, $24,25(OH)_2D_3$, and $1,25(OH)_2D_3$ are shown. In the bottom panel, the elution pattern of the chloroform extract from keratinocytes incubated for one hour with 0.1 μCi of $[^3H]$ $25(OH)D_3$ is shown. Only the tracing from the radioactivity monitor is depicted. The presumptive $24,25(OH)_2D_3$ and $1,25(OH)_2D_3$ peaks are identified.

experiment, although in other experiments a substantial peak eluting in the position of $1,24,25(OH)_3D_3$ was observed (see description below). This peak also was periodate sensitive. The unidentified metabolites are referred to as peaks 22, 24, 27, and 28 according to their elution time in minutes in this chromatographic system. Peaks 27 and 28 elute in the positions expected for 23 or 24 keto, $1,25(OH)_2D_3$ and $1,23,25(OH)_3D_3$, respectively (Napoli et al, 1983).

The amount of these metabolites recovered from the cells and the medium depended on the duration of the incubation as shown in Figure 4 A-F. A metabolite in peak 22 rapidly appeared in the medium, reaching a peak at 15 minutes, followed by an exponential decay over the next 24 hours. A second metabolite may have appeared in peak 22, increasing in the cell more slowly and reaching a maximum level at 4 hours before disappearing by 24 hours. The metabolite $1,25(OH)_2D_3$ was found in greatest concentration 1 hour after the addition of $25(OH)D_3$. Very little $1,25(OH)_2D_3$ was released into the medium. The amount of peak 24 in the cells and in the medium increased in parallel (although the cells contained more peak 24 than did the medium at all times), reaching a maximum concentration 8 hours after the addition of $25(OH)D_3$. Peak 27 also increased in parallel in both cells and medium, reaching a maximum concentration 4 hours after the addition of $25(OH)D_3$. Unlike peak 24, peak 27 was found in greater amounts in the medium than in the cells. Peak 28 appeared more quickly in the cells than in the medium, although maximum levels were reached in both by 4 hours; after 4 hours peak 28 disappeared more quickly from cells than from medium. Little $24,25(OH)_2D_3$ or $1,24,25(OH)_3D_3$ was formed in this experiment (data not shown). Little radioactivity was found in the aqueous extract of the cells at any time. Instead, the water soluble metabolites appeared in the medium where they increased in amount, at least through 16 hours of incubation.

The data in Figure 4b indicate that $1,25(OH)_2D_3$ was the major metabolite of $25(OH)D_3$ produced by these keratinocytes during the first 4 hours of incubation. Maximal levels of $1,25(OH)_2D_3$ (at 1 hour) were 190 fmol/well compared to 35 fmol/well of peak 22 (at 4 hours), 47 fmol/well of peak 24 (at 8 hours), 55 fmol/well of peak 27 (at 4 hours), and 90 fmol/well of peak 28 (at 4 hours).

The substrate dependence of $1,25(OH)_2D_3$ production was determined using a 1 hour incubation and a total $25(OH)D_3$ concentration from 0.65 x 10^{-9} M (tracer only) to 500 x 10^{-9} M. The results analyzed by an Eadie-Hofstee plot (Fig. 5) indicated an apparent K_m for $25(OH)D_3$ of 5.4 x 10^{-8} M. This value is only approximate because, as shown in Figure 2, the actual substrate ($25(OH)D_3$) concentration available to the intracellular 1α-hydroxylase increased during the 1 hour incubation and thus is less than the total substrate concentration upon which this calculation was made. Furthermore, a portion of the $1,25(OH)_2D_3$ produced may have been catabolized during the 1 hour incubation so that the amount recovered may have been less than the amount formed (see below). The limited amount of other metabolites

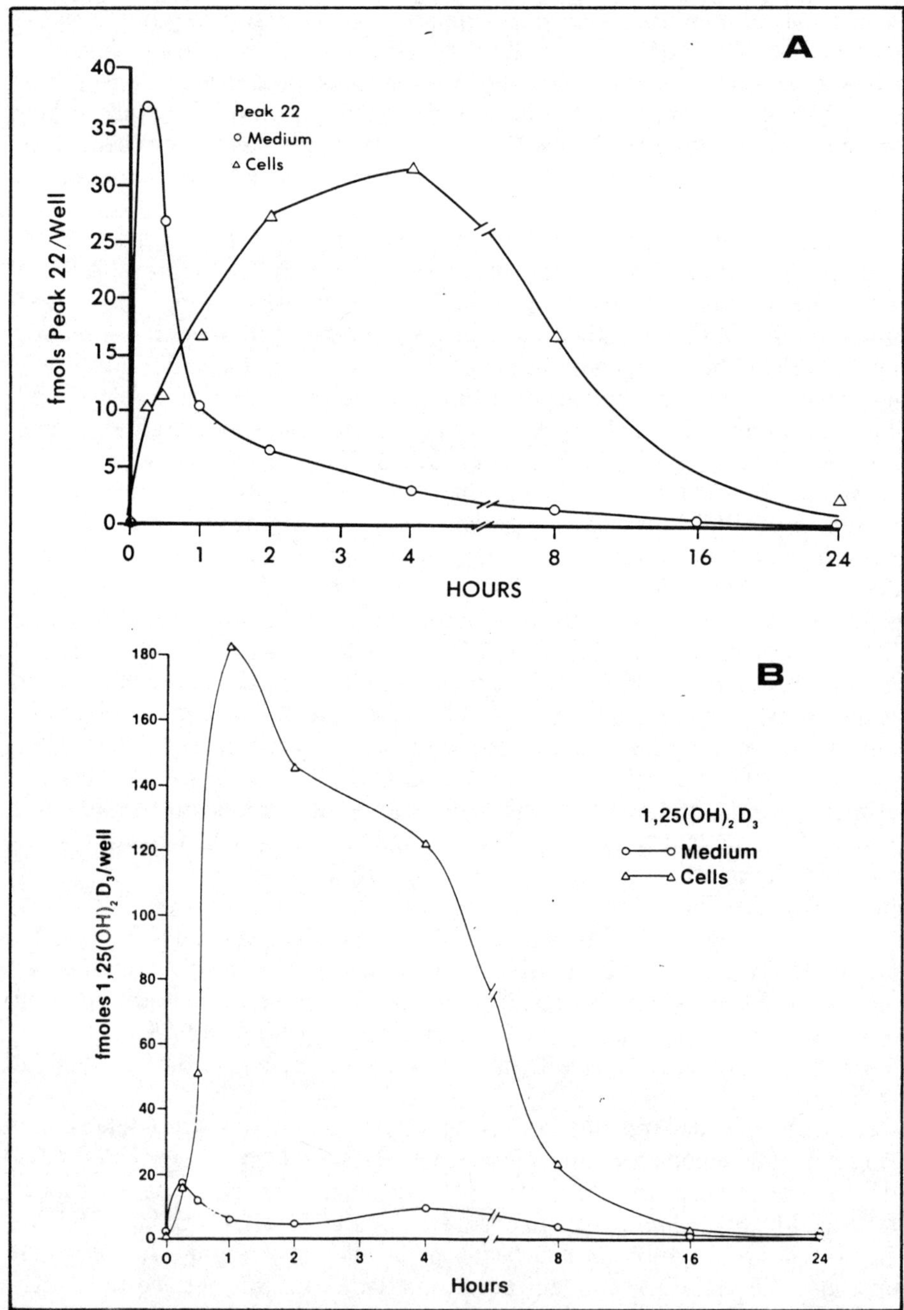

Figure 4. (A-F) The effect of time of incubation on the recovery of the major metabolites of $25(OH)D_3$ in cells and medium following the addition of 1600 fmoles $[^3H]$ $25(OH)D_3$ to cultured keratinocytes. This experiment illustrates the production and secretion into the medium of the major metabolites of $25(OH)D_3$ by keratinocytes. Each point is the mean of duplicate dishes.

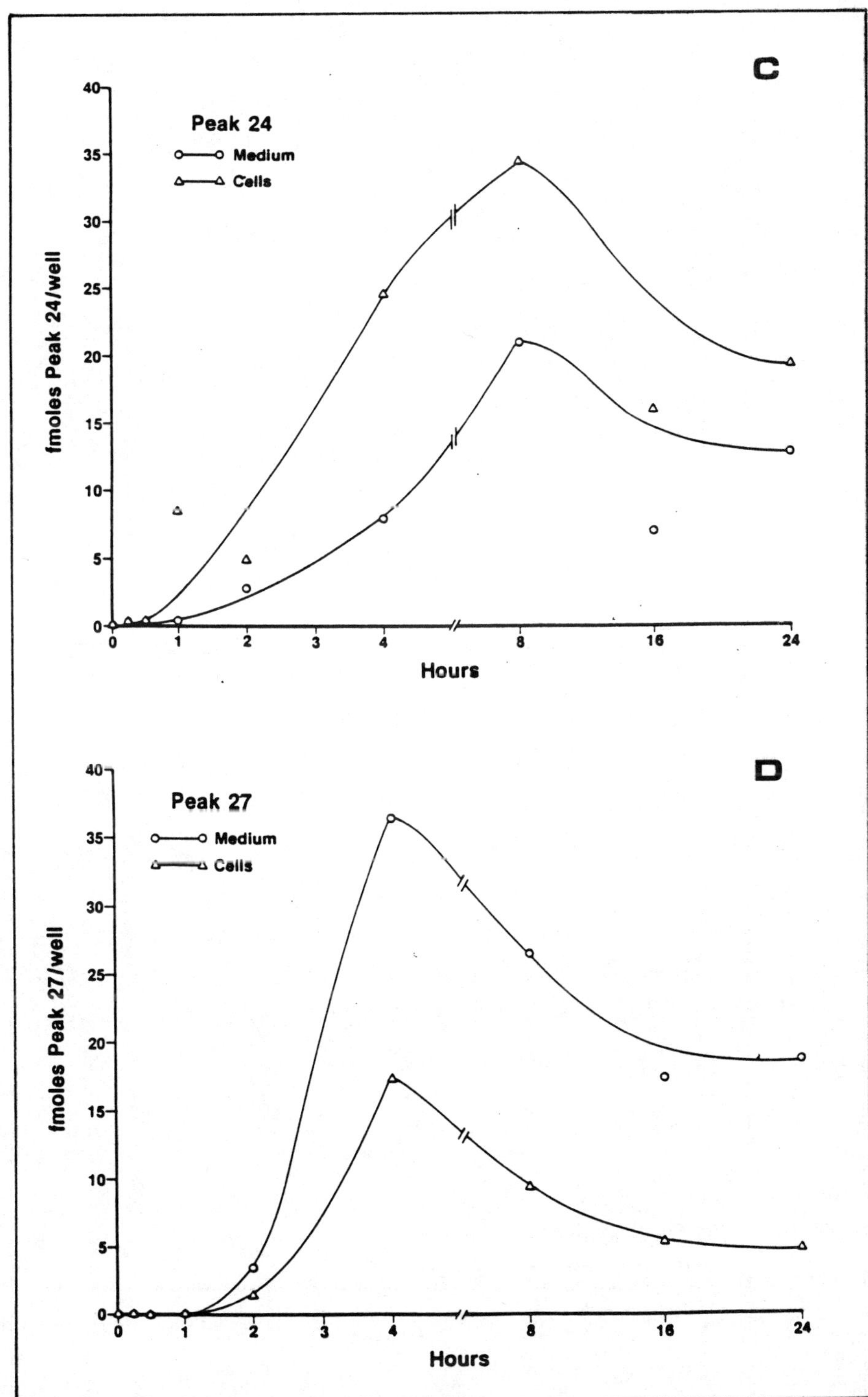
C
Peak 24
Medium
Cells
fmoles Peak 24/well
40
35
30
25
20
15
10
5
0
0
1
2
3
4
8
16
24
Hours
D
Peak 27
Medium
Cells
fmoles Peak 27/well
40
35
30
25
20
15
10
5
0
0
1
2
3
4
8
16
24
Hours

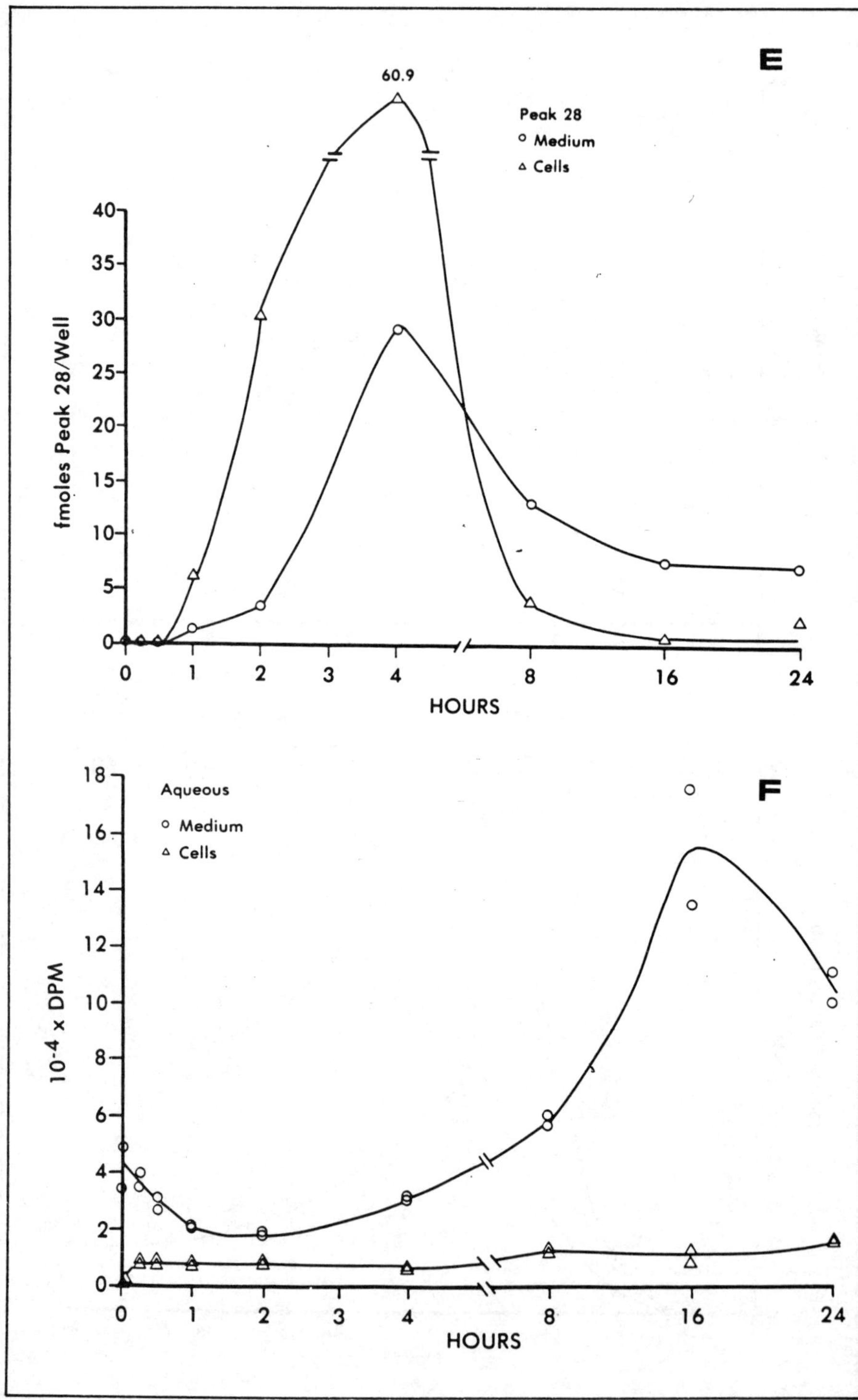
E
60.9
Peak 28
○ Medium
△ Cells
fmoles Peak 28/Well
0
5
10
15
20
25
30
35
40
0
1
2
3
4
8
16
24
HOURS
F
Aqueous
○ Medium
△ Cells
10-4 x DPM
0
2
4
6
8
10
12
14
16
18
0
1
2
3
4
8
16
24
HOURS

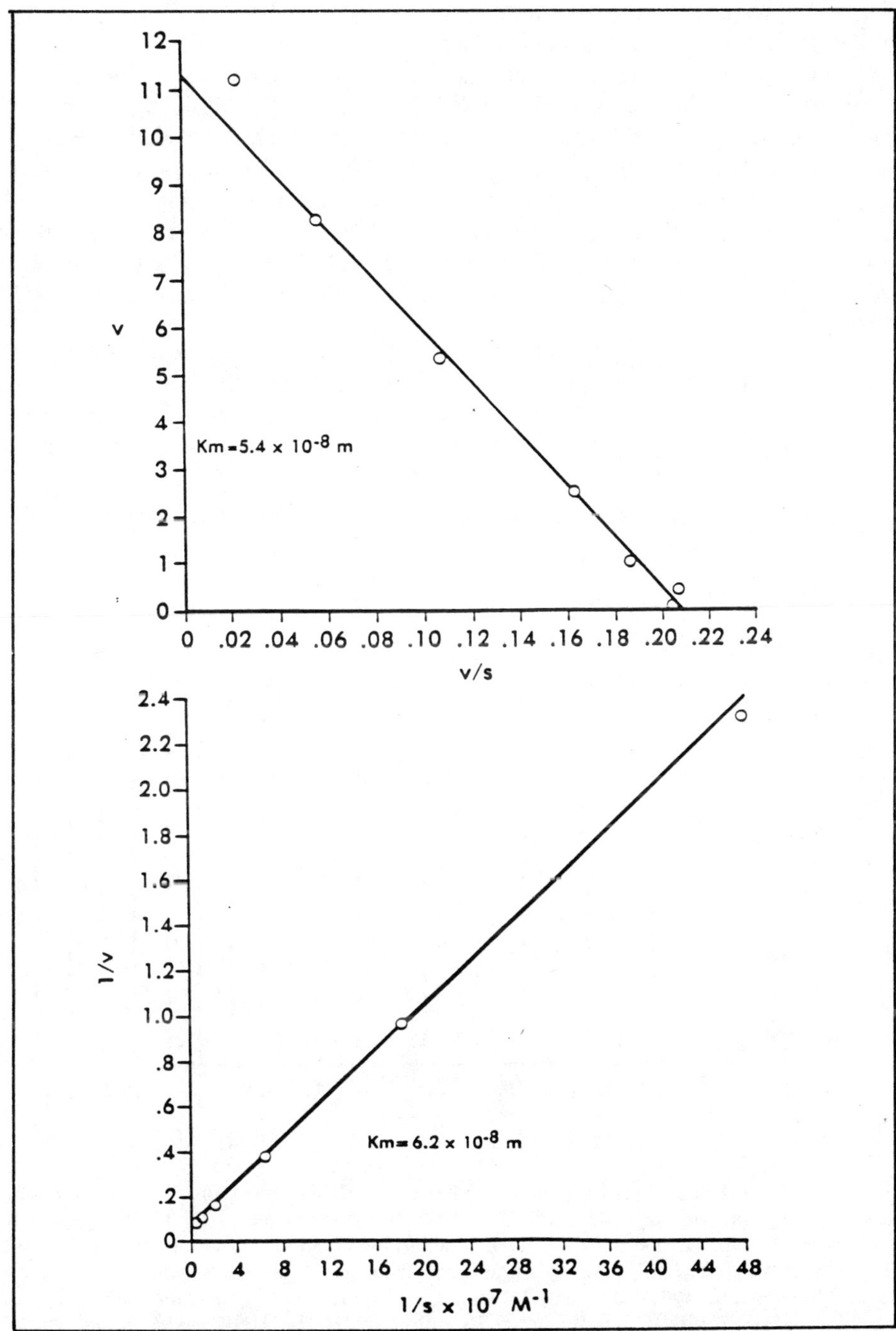

Figure 5. The Lineweaver-Burk (A) and Eadie-Hofstee(B) analyses of $1,25(OH)_2D_3$ production by cultured keratinocytes incubated for one hour at various $25(OH)D_3$ concentrations. The velocity (V) units are pmol $1,25(OH)_2D_3$ per h/dish; the substrate(S) units are 10^{-7} M for A and 10^{-91} M for B. Each point is the mean of duplicate dishes.

formed during the 1 hour incubation suggests that catabolism of the $1,25(OH)_2D_3$ was not substantial, however.

Regulation of $25(OH)D_3$ metabolism was explored. Increasing the calcium concentration in the medium from 0 (calcium free) to 4 mM during incubation had little effect on $25(OH)D_3$ metabolism (not shown). Likewise, 8-Br-cAMP (0 to 10^{-6} M) or dibutyryl cAMP (0 to 10^{-4} M) had little effect (not shown). However, in the experiments described below (Fig. 6), 1 mM isobutylmethylxanthine (IBMX) increased $1,25(OH)_2D_3$ levels from 94 to 139 fmol/well when added concurrently with $[^3H]$ $25(OH)D_3$, and from 92 to 203 fmol/well when added 4 hour prior to a 4 hour incubation with $[^3H]$ $25(OH)D_3$.

PTH increased $1,25(OH)_2D_3$ levels after a 4 hour incubation with $[^3H]$ $25(OH)D_3$ (Fig.6). In the absence of IBMX, increasing the PTH concentra-

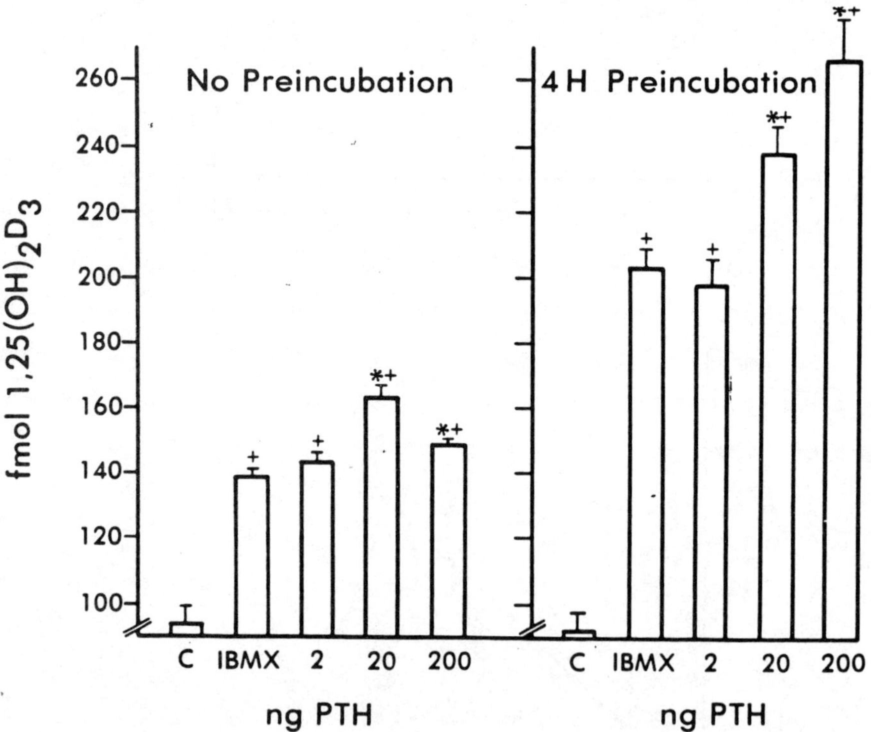

Figure 6. The effect of 1 mM isobutylmethylxanthine (IBMX) alone or in combination with various concentrations of bovine PTH 1-34. In the study depicted in the left panel (no preincubation), the IBMX and PTH were added to the keratinocytes at the same time as the $[^3H]$ $25(OH)D_3$ (625 fmol) to begin a four hour incubation. In the study depicted in the right panel (4h preincubation) IBMX and PTH were added to the keratinocytes four hours before the addition of $[^3H]$ $25(OH)D_3$. Following the addition of $[^3H]$ $25(OH)D_3$ and a second dose of PTH, the cultures were incubated for four more hours. The control cells received only vehicle (20 μl 10 mM acetic acid in 0.1% bovine serum albumin). The error bars enclose the mean $\pm$ range of duplicate dishes. The + indicates results that are significantly ($p < 0.05$) greater than control (C). The * indicates results that are significantly ($p < 0.05$) greater than IBMX alone.

tion from 0 to 20 ng/ml raised $1,25(OH)_2D_3$ levels 33% from 54 to 72 fmol/well. Higher concentrations of PTH (200 ng/ml) were less effective. When PTH was added with IBMX at the start of the 4 hour incubation with $[^3H]$ $25(OH)D_3$, it enhanced the stimulatory effect of IBMX; again at the optimal concentration of 20 ng/ml, 163 fmol $1,25(OH)_2D_3$/well were recovered in comparison to 139 fmol/well in the presence of IBMX alone. When the cells were incubated for 4 hours with IBMX and PTH before $[^3H]$ $25(OH)D_3$ was added, the 200 ng/ml PTH concentration was optimal in increasing $1,25(OH)_2D_3$ levels, raising them from 203 fmol/well in the absence of PTH to 266 fmol/well in the presence of PTH. In this experiment the recovery of 266 fmol of $1,25(OH)_2D_3$ represented 43% of the initial substrate added (625 fmol of $[^3H]$ $25(OH)D_3$).

The effects of IBMX and PTH on the levels of $1,25(OH)_2D_3$ were associated with marked changes in the levels of other metabolites. In particular, the amount of peak 28 showed a reciprocal relationship to the amount of $1,25(OH)_2D_3$ recovered during these 4 hour incubations with $[^3H]$ $25(OH)D_3$. The combination of PTH and IBMX markedly inhibited the appearance of peak 28. Thus, these effects on $1,25(OH)_2D_3$ levels are consistent either with a stimulation by PTH and IBMX of $1,25(OH)_2D_3$ production or with an inhibition by PTH and IBMX of $1,25(OH)_2D_3$ catabolism to other metabolites (such as peak 28).

To determine how PTH and IBMX raise $1,25(OH)_2D_3$ levels in keratinocytes we performed two types of experiments. First we determined the effects of PTH and IBMX on $25(OH)D_3$ metabolism at three different times of incubation: 1 hour, 4 hour, and 8 hour. The results for all major metabolites of $25(OH)D_3$ formed in this experiment are shown in Table 2. In this experiment metabolites eluting in the position of $24,25(OH)_2D_3$ and $1,24,25(OH)_3D_3$ were observed. At 1 hour of incubation, $1,25(OH)_2D_3$ was the major metabolite formed, and its production was stimulated from 70 to 94 fmol/well by 20 ng/ml PTH and from 70 to 147 fmol/well by IBMX. No detectable peak 28 was formed by 1 hour. Levels of $24,25(OH)_2D_3$ were little affected by PTH but were abolished by IBMX, whereas $1,24,25(OH)_3D_3$ levels were increased from 2.3 to 4.7 fmol/well by PTH and to 11 fmol/well by IBMX. After a 4 hour incubation with $[^3H]$ $25(OH)D_3$, peak 28 and $1,24,25(OH)_3D_3$ were the major metabolites recovered (113 fmol/well and 89 fmol/well, respectively) although substantial amounts of $1,25(OH)_2D_3$ (54 fmol/well) were still present. PTH increased the levels of both $1,25(OH)_2D_3$ (to 79 fmol/well) and $1,24,25(OH)_3D_3$ (to 110 fmol/well) with little effect on the other metabolites. IBMX increased $1,25(OH)_2D_3$ levels to 231 fmol/well (a 427% increase over controls), reduced $24,25(OH)_2D_3$ levels from 23 to 2 fmol/well, reduced peak 28 levels from 113 to 14 fmol/well, and reduced $1,24,25(OH)_3D_3$ levels from 89 to 46 fmol/well. After an 8 hour incubation, most of the radioactive label was in the form of water soluble metabolites (263 fmol equivalents/well). IBMX markedly inhibited the production of the water soluble metabolites (reducing them from 263 to 113 fmol equivalents/well), which resulted in increased levels of the chloroform extractable

Table 2. Time course of PTH and IBMX effects on 25(OH)D_3 metabolism.

		fmol recovered						
ng/ml PTH	mM IBMX	24,25$(OH)_2D_3$	Peak 22	1,25$(OH)_2D_3$	Peak 24	Peak 28	1,24,25$(OH)_3D_3$	Aqueous
1 Hour incubation								
0	0	4.7 ± .9	3.1 ± .1	70 ± 3	1.6 ± .3	—	2.3 ± .1	7.2 ± .3
20	0	3.1 ± .6	1.9 ± .6	94 ± 9	1.3 ± .1	—	4.7 ± .9	11 ± .9
0	1	—	5.0 ± 1.9	147 ± 9	1.6 ± .9	—	11 ± 4	13 ± .3
20	1	—	1.9 ± 1.9	144 ± 4	2.5 ± 1.2	—	12 ± 4	18 ± 2
4 Hour incubation								
0	0	23 ± .6	28 ± 3	54 ± .6	36 ± 3	113 ± 6	89 ± 1	49 ± 1
20	0	23 ± 6	31 ± 6	79 ± 8	34 ± 6	102 ± .3	110 ± 2	53 ± 3
0	1	2 ± 1	23 ± 4	231 ± 5	10 ± .9	13 ± 5	46 ± 12	36 ± 2
20	1	1.3 ± .6	33 ± 3	274 ± 9	12 ± 2	12 ± 1	50 ± 3	38 ± 3
8 Hour incubation								
0	0	17 ± 1	11 ± .1	28 ± .6	29 ± 2	33 ± 2	14 ± .9	263 ± 8
20	0	21 ± 2	23 ± 6	34 ± 5	22 ± 4	43 ± 3	11 ± 3	275 ± 10
0	1	32 ± 8	61 ± 6	28 ± .6	25 ± 9	91 ± 11	21 ± 1	113 ± 13
20	1	25 ± 6	52 ± 3	48 ± 6	33 ± .3	106 ± 34	62 ± 18	102 ± 20

In this experiment the cells were preincubated with the indicated concentrations of PTH and IBMX for 4 h prior to the addition of 625 fmol [^3H] 25(OH)D_3. The incubation time refers to the duration of the incubation following the addition. Duplicate results are expressed as fmol/well ± range.

metabolites, including peak 28. PTH in the presence of IBMX selectively increased the recovery of $1,25(OH)_2D_3$ (from 28 to 48 fmol/well) and $1,24,25(OH)_3D_3$ (from 21 to 62 fmol/well) in the cells incubated with [^3H] $25(OH)D_3$ for 8 hours.

The second type of experiment was to directly assess the effects of IBMX and PTH on the metabolism of $1,25(OH)_2D_3$. We added radioactively labeled $1,25(OH)_2D_3$ to keratinocytes following a 4 hour incubation with 20 ng/ml PTH or 1 mM IBMX for an additional 4 hour incubation. The results of this experiment are shown in Table 3. In this experiment, only 29% of the added [^3H] $1,25(OH)_2D_3$ was recovered as $1,25(OH)_2D_3$ in the absence of PTH and IBMX. Thirty percent appeared in peak 28 and 22% appeared as $1,24,25(OH)_3D_3$. PTH had little effect on the metabolism of $1,25(OH)_2D_3$ in the absence of IBMX. In contrast, the inclusion of IBMX in the medium resulted in substantially less metabolism of $1,25(OH)_2D_3$. Forty-seven percent of the added [^3H] $1,25(OH)_2D_3$ was recovered as $1,25(OH)_2D_3$ in the presence of IBMX; the percent recovered in peak 28 fell from 31% to 10.5%, whereas the amount recovered as $1,24,25(OH)_3D_3$ was not changed. The addition of PTH in the presence of IBMX further increased the recovery of $1,25(OH)_2D_3$ to 59% and reduced the conversion of $1,25(OH)_2D_3$ to peak 28 (6.6%) and $1,24,25(OH)_3D_3$ (14%). These results suggest that IBMX raises $1,25(OH)_2D_3$ levels in keratinocytes by retarding the catabolism of $1,25(OH)_2D_3$. PTH, on the other hand, appears to increase $1,25(OH)_2D_3$ (and $1,24,25(OH)_3D_3$) levels primarily by stimulating the 1 α-hydroxylase, but also has a synergistic effect with IBMX to retard $1,25(OH)_2D_3$ catabolism.

We then examined the ability of exogenous $1,25(OH)_2D_3$ to regulate its own production. When $1,25(OH)_2D_3$ was added to keratinocyte cultures 16 hours prior to the addition of 676 fmol [^3H] $25(OH)D_3$, the recovery of [^3H] $1,25(OH)_2D_3$ was markedly reduced (Fig. 7). At the lowest dose tested (10^{-12} M), the recovery of [^3H] $1,25(OH)_2D_3$ (75 fmol/well) was 83% of that found in the absence of exogenous $1,25(OH)_2D_3$ (90 fmol/well), and at 10^{-11} M $1,25(OH)_2D_3$, the recovery of [^3H] $24,25(OH)_2D_3$ was reduced by 50%. In contrast, the recovery of [^3H] $24,25(OH)_2D_3$ was increased from 10 to 28

Table 3. Effect of PTH and IBMX on $1,25(OH)_2D_3$ metabolism.

		% recovered			
PTH (ng/ml)	IBMX (mM)	$1,25(OH)_2D_3$	Peak 28	$1,24,25(OH)_3D_3$	Aqueous
0	0	28.9 ± 1.5	30.7 ± 1.2	21.8 ± 0.8	8.5 ± 0.5
20	0	28.5 ± 4.0	28.2 ± 1.0	21.2 ± 0.7	7.3 ± 2.9
0	1	46.9 ± 3.3	10.5 ± 0.9	21.0 ± 4.7	4.6 ± 0.3
20	1	58.5 ± 8.0	6.6 ± 0.4	14.4 ± 1.0	4.6 ± 0.3

In this experiment the cells were incubated with the indicated concentrations of PTH and/or IBMX for 4 h at which point 0.05μ Ci [^3H] $1,25(OH)_2D$ was added for an additional 4 h incubation. The data expressed as percent of the label recovered in the indicated peak following HPLC of the chloroform extract or in the aqueous extract. Each incubation was performed in duplicate, and the mean ± range of duplicates are shown.

fmol/well by the presence of 10^{-11} M $1,25(OH)_2D_3$. As the concentration of exogenous $1,25(OH)_2D_3$ increased, the recovery of $[^3H]$ $1,25(OH)_2D_3$ fell to essentially zero, while the recovery of $[^3H]$ $24,25(OH)_2D_3$ continued to increase, reaching 48 fmol/well at 10^{-8} M $1,25(OH)_2D_3$. The recovery of $[^3H]$ $1,24,25(OH)_3D_3$ was increased (from 11 to 27 fmol/well) by increasing $1,25(OH)_2D_3$ from 0 to 10^{-10} M, but at higher concentrations of exogenous $1,25(OH)_2D_3$, the recovery of this metabolite fell in parallel with the fall in recovery of $[^3H]$ $1,25(OH)_2D_3$. The amount of radioactivity recovered from the aqueous extract (38 fmol equivalents/well in controls) was increased by concentrations of exogenous $1,25(OH)_2D_3$ greater than 10^{-12} M, reaching a level of 98 fmol equivalents/well at 10^{-8} M $1,25(OH)_2D_3$. Similar results

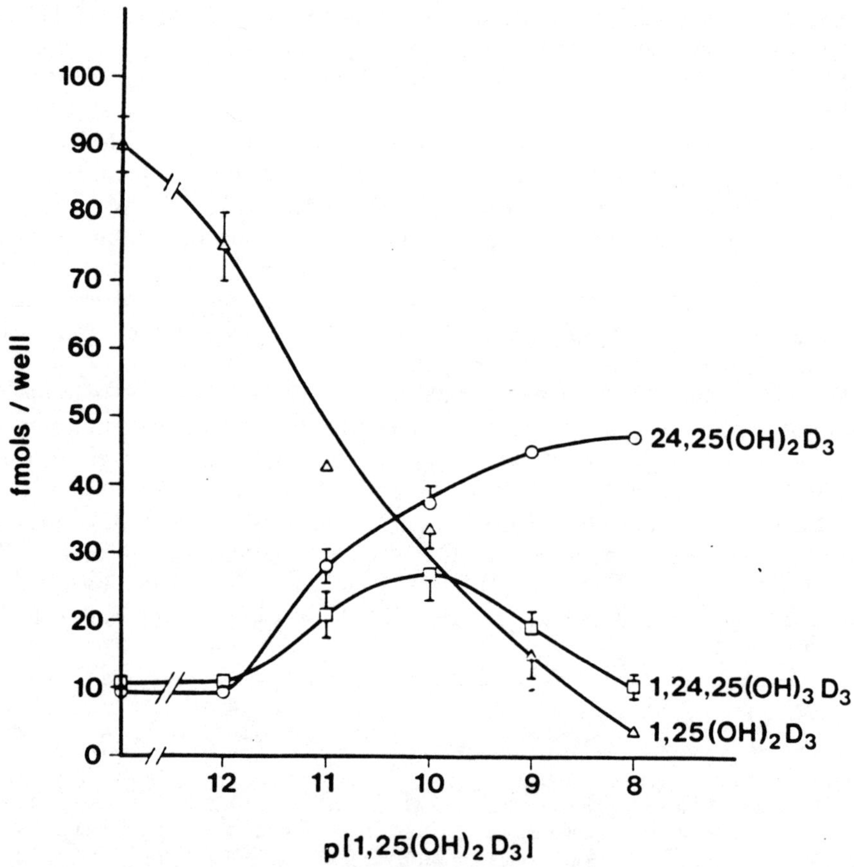

Figure 7. The effect of $1,25(OH)_2D_3$ on $25(OH)D_3$ metabolism by keratinocytes. Keratinocytes were incubated with nonradioactively labeled $1,25(OH)_2D_3$ at the indicated concentrations for 16 hours. The medium was removed, the cells were washed once with fresh medium and were then incubated with $[^3H]$ $25(OH)D_3$ for one hour. The production of $1,25(OH)_2D_3$ (▽), $24,25(OH)_2D_3$ (○), and $1,24,25(OH)_3D_3$ (□) is shown. Each point is the mean value range of duplicate determinations.

were obtained when a 4 hour incubation with [^{3}H] 25(OH)D_3 was used except that 10^{-12} M exogenous 1,25$(OH)_2D_3$ was sufficient to reduce recoverable [^{3}H] 1,25$(OH)_2D_3$ to 50% of control (data not shown). None of these changes induced by 1,25$(OH)_2D_3$ was found when actinomycin D (2 μg/ml) was added to the medium at the time 1,25$(OH)_2D_3$ was added (data not shown). Actinomycin D markedly reduced [^{3}H] 1,25$(OH)_2D_3$ recovery (to 11 fmol/well) in all cultures regardless of the presence of exogenous 1,25$(OH)_2D_3$. In a parallel experiment, incubation of keratinocytes with these same concentrations of 1,25$(OH)_2D_3$ for 16 hours did not significantly decrease the protein or DNA content of the cultures or alter the extent to which the cells excluded trypan blue.

Because of the possibility that exogenous 1,25$(OH)_2D_3$ might affect its own catabolism as well as production, we incubated [^{3}H] 1,25$(OH)_2D_3$ for 1 hour with control cells or with cells previously exposed to 10^{-8} M 1,25$(OH)_2D_3$ for 16 hours. The results are shown in Table 4. Even without preincubation with 1,25$(OH)_2D_3$, only 61% of the added [^{3}H] 1,25$(OH)_2D_3$ could be recovered as 1,25$(OH)_2D_3$ after 1 hour incubation; 10.4% of the radioactive label appeared as peak 28 and 13% as 1,24,25$(OH)_2D_3$. Preincubation with nonlabeled 1,25$(OH)_2D_3$ markedly enhanced the ability of the cells to metabolize [^{3}H] 1,25$(OH)_2D_3$ to both peak 28 (28%) and 1,24,25$(OH)_3D_3$ (19%); only 20% of the added [^{3}H] 1,25$(OH)_2D_3$ could be recovered as 1,25$(OH)_2D_3$. In both control and 1,25$(OH)_2D_3$-treated cells, small amounts (less than 5%) of the radioactive label appeared in peaks 22,24, and 27 (not shown).

Identification of 1,25(OH)$_2$D$_3$

To produce adequate amounts of the presumptive 1,25$(OH)_2D_3$ for structural identification, keratinocytes were grown to confluence on twenty 10 cm diameter plates, then serum deprived for 36 hours. The medium was then supplemented with 1 mM IBMX and 1 μg/ml bovine parathyroid hormone 1-34 (to maximize 1,25$(OH)_2D_3$ production) for 4 hours before adding ^{3}H-25(OH)D_3 to final concentration of 500×10^{-9} M (specific activity 2 Ci/mmol). The incubation was continued for an additional 2 hours; cells and

Table 4. Metabolism of [^{3}H] 1,25$(OH)_2D_3$ by keratinocytes as affected by prior incubation with 10^{-8} M nonradioactive 1,25$(OH)_2D_3$.

	% recovered			
	1,25$(OH)_2D_3$	Peak 28	1,24,25$(OH)_3D_3$	Aqueous
Control	60.7 ± 2.3	10.4 ± 0.5	12.9 ± 0.7	8.0 ± 0.8
1,25$(OH)_2D_3$	20.5 ± 4.4	27.8 ± 2.1	19.4 ± 2.5	27.0 ± 1.9

0.05 μCi [^{3}H] 1,25$(OH)_2D_3$ was incubated for 1 h with keratinocytes preincubated for 16 h with either vehicle (control) or 10^{-8} M nonradioactive 1,25$(OH)_2D_3$. The data are shown as percent recovery of radioactivity in the various peaks following HPLC of the chloroform extract or in the aqueous extract. Each incubation was performed in duplicate, and the data are given as mean ± range or duplicates.

medium were then extracted as above, and the chloroform extract was chromatographed on 7.8 mm x 30 cm μ Porasil column, eluting with 90:10 hexane:isopropanol. The presumptive 1,25$(OH)_2D_3$ peak (peak 23, the major metabolite produced under these conditions) was further purified using three chromatographic systems. The reverse-phase system used a 3.9 mm x 30 cm C18 μ-Bondapak column eluted with 75:25 MeOH:H_2O. The two straight-phase systems each used a 4.6 mm $\times$ 25 cm Zorbax Sil column eluted either with a 97:3-90:10 hexane:isopropanol nonlinear gradient or with a 96:4 dichloromethane:isopropanol isocratic system. Under these conditions, 5.6% of the substrate was converted to a metabolite eluting in the 1,25$(OH)_2D_3$ position, and 70%-80% of this metabolite was recovered from each of the chromatographic steps in the same position as authentic 1,25$(OH)_2D_3$

The chromatographic elution pattern of the presumptive 1,25$(OH)_2D_3$ in two additional chromatographic systems is shown in Figures 8a and 9a. Figures 8b and 9b show the elution pattern of authentic 1,25$(OH)_2D_3$ (a gift from M.R. Uskokovic, Hoffmann LaRoche, Nutley, NJ) compared to the elution pattern of keratinocyte peak 23. The chromatographic system used to obtain the data in Figure 8 was a reverse-phase system (a C18 μ-Bondapak column eluted with 75:15 MeOH:H_2O) The UV and radioactivity monitor recordings are shown. Peak 23 eluted in a position identical to the authentic 1,25$(OH)_2D_3$ standard. In Figure 9, data are shown for a straight-phase system (a Zorbax Sil column eluted with 96:4 dichloromethane:isopropanol). Again, peak 23 elutes in the same position as authentic 1,25$(OH)_2D_3$.

Peak 23 was then assayed for its ability to displace $[^3H]$ 1,25$(OH)_2D_3$ from the chick intestinal cytosol receptor using the method described by Eisman et al. The results are shown in Figure 10. Based on the specific activity of the 25$(OH)D_3$ used as substrate to produce peak 23, this metabolite is as effective as authentic 1,25$(OH)_2D_3$ in displacing tracer 1,25$(OH)_2D_3$ from the receptor.

Mass spectrometric determinations of peak 23 (Fig. 11, bottom) and authentic 1,25$(OH)_2D_3$ (Fig. 11, top) were then performed with a Kratos MS-12 mass spectrometer equipped with an Incos Data System, using a direct insertion probe at temperatures 100°C to 200°C above ambient and a scan rate of 30 sec/decade. The molecular ion is seen at m/z 416, and fragment ions are seen at m/z 398,380, and 362 (indicating losses of 1,2, and 3 molecules of water, respectively), and m/z 383, 365, and 347 (indicating the additional loss of CH_3). Other characteristic ions in the spectrum include m/z 287,269, and 251 (indicating a loss of the 8-membered side chain and losses of 1 and 2 water molecules, respectively), and m/z 152 and 134 (indicating fission between C7 and C8, and the loss of 1 water molecule, respectively). The latter fragmentations are diagnostic of the intact nuclear triene system of vitamin D metabolites. The essential features of the mass spectrum of peak 23 are identical to those of the authentic 1,25$(OH)_2D_3$ standard that was analyzed at the same time, and to the mass spectrum that Holick (1971) used to originally identify the structure of 1,25$(OH)_2D_3$.

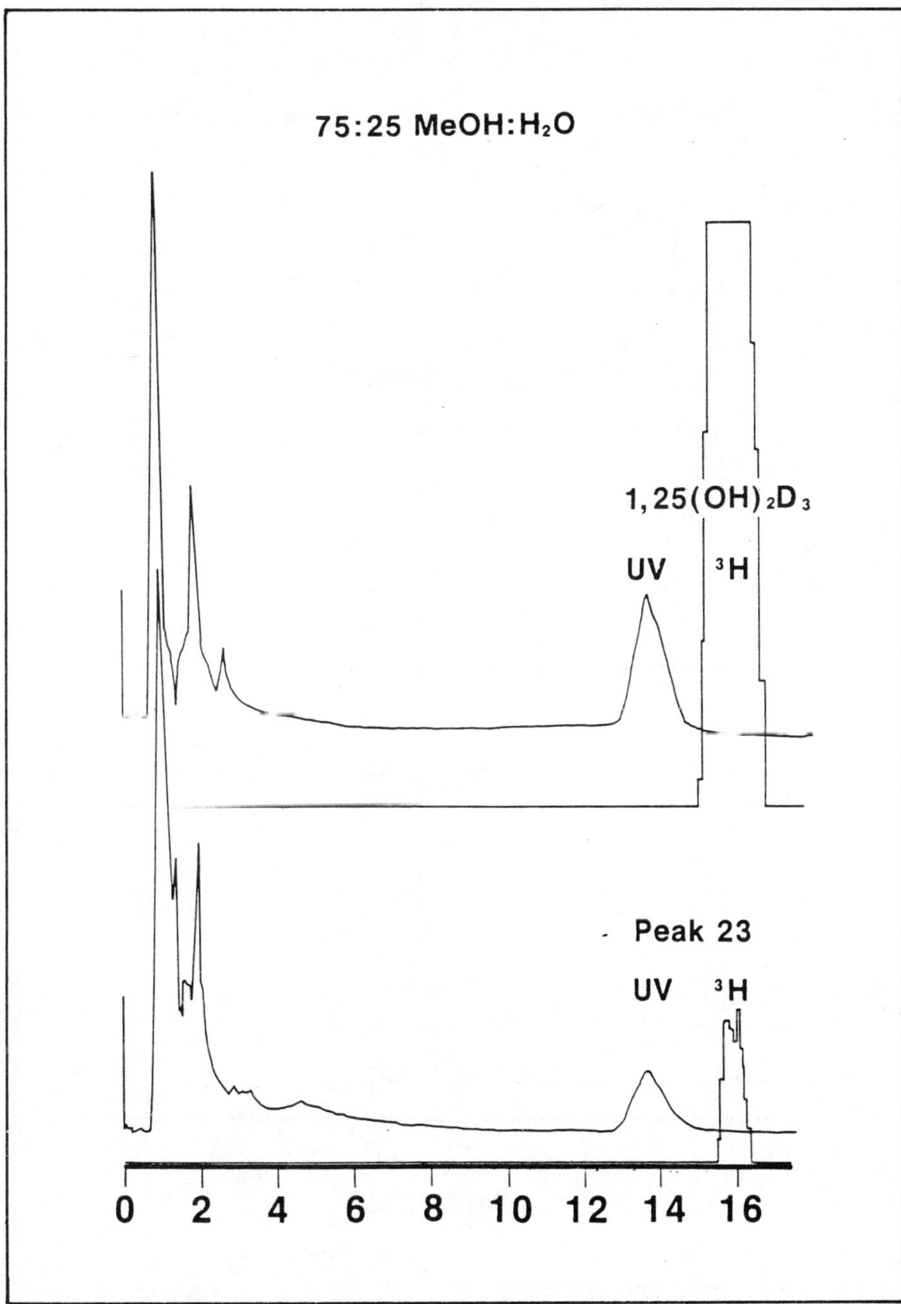

Figure 8. Elution pattern of radioactive and nonradioactive $1,25(OH)_2D_3$ standards (panel A) compared with that of purified peak 23 (panel B). A 3.9 mm × 30 cm C_{18} μ-Bondapak column eluted with 75:25 methanol:water at 2 ml/min. was used. The presumptive $1,25(OH)_2D_3$ was generated by keratinocytes incubated with 500×10^{-9} M [^{3}H] $25(OH)D_3$ (sp. activity, 2 Ci/mol) and initially purified by using a 7.8 cm × 30 cm μ Porasil column prior to this chromatographic analysis.

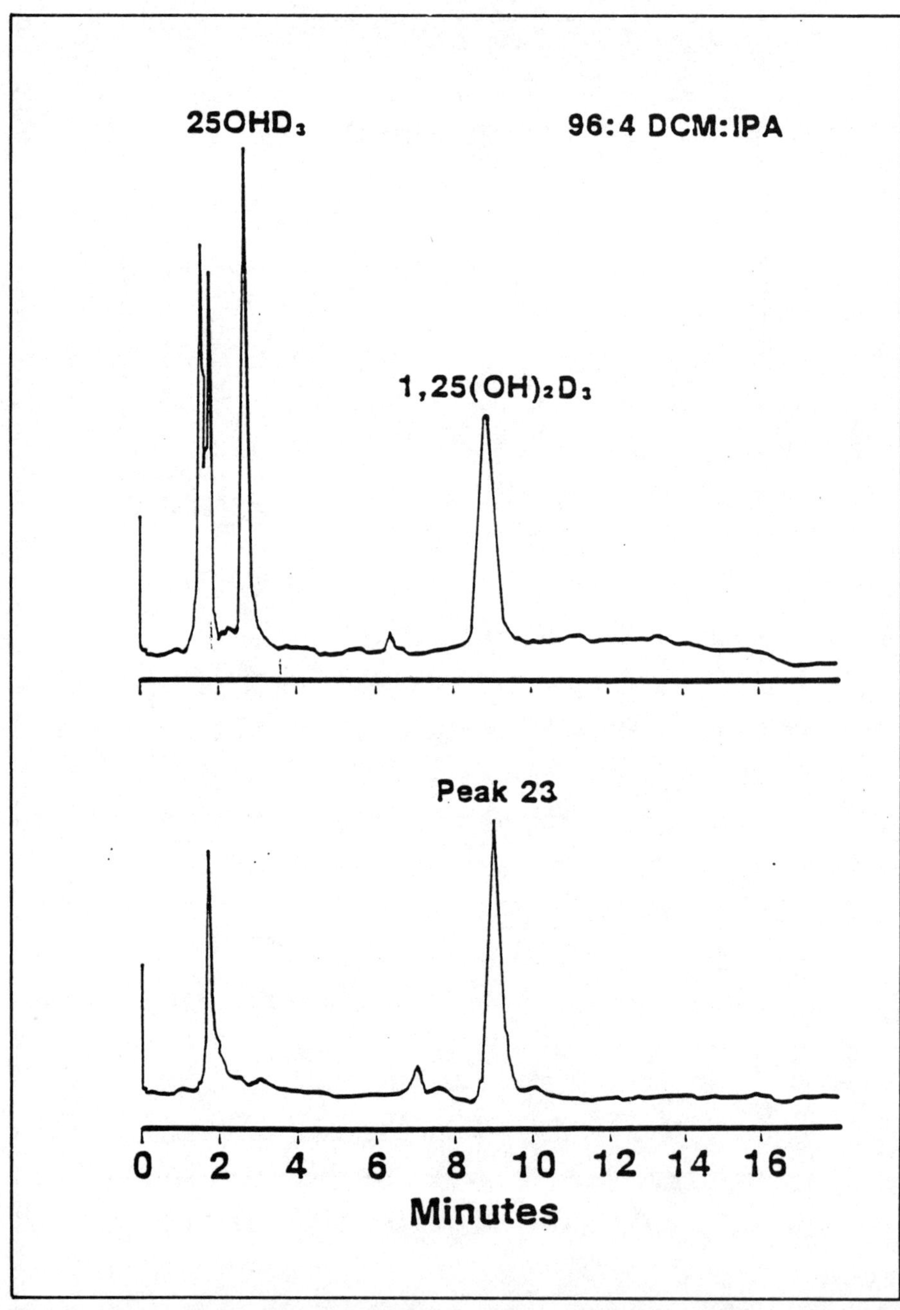

Figure 9. Elution pattern of $25(OH)D_3$ and $1,25(OH)_2D_3$ standards (panel A) compared with that of purified peak 23 (panel B). A 4.6 mm × 25 cm Zorbax Sil column eluted with 96:4 dichloromethane-2 propanol at 2 ml/min was used. Only the UV monitor trace is shown since dichloromethane quenched the radioactivity below the limits of detection under these conditions. The presumptive $1,25(OH)_2D_3$ was prepared as described in Figure 8 prior to this chromatographic analysis.

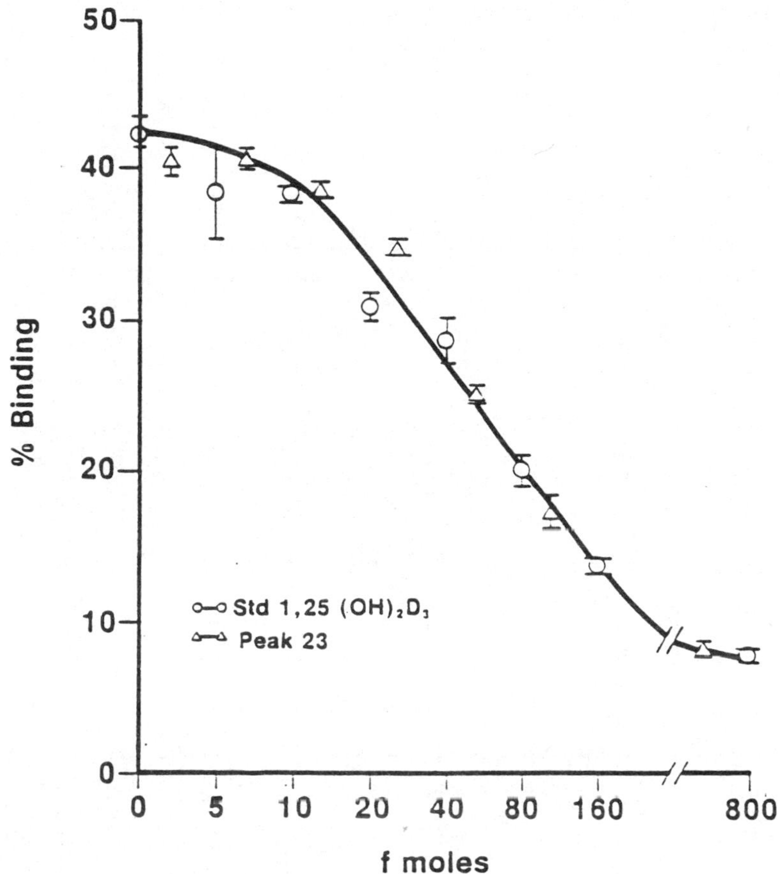

Figure 10. Comparison of the ability of peak 23 (▽) and authentic $1,25(OH)_2D_3$ (○) to displace labeled $1,25(OH)_2D_3$ from the chick intestinal cytosol receptor. The amount of peak 23 that was added was determined by the specific activity of the initial substrate ($25(OH)D_3$) and the radioactivity in the final purified product (peak 23). Peak 23 and authentic $1,25(OH)_2D_3$ have comparable affinity for the intestinal receptor. Error bars enclose mean and range of duplicate determinations.

These results indicate that human foreskin keratinocytes produce $1,25(OH)_2D_3$.

Vitamin D metabolism in liver and kidney

a. Introduction. 25 hydroxy vitamin D_3 is produced in the liver. Its concentration in the serum is reduced by the administration of $1,25(OH)_2D_3$. It had not been known whether $1,25(OH)_2D_3$ inhibited the synthesis of $25(OH)D_3$ in the liver, or whether $1,25(OH)_2D_3$ increased the metabolic turnover or catabolism of $25(OH)D_3$. Therefore, we evaluated the effect of $1,25(OH)_2D_3$ on the catabolism of $25(OH)D_3$, $1,25(OH)_2D_3$ and $24,25(OH)_2D_3$ in rat liver and kidney homogenates.

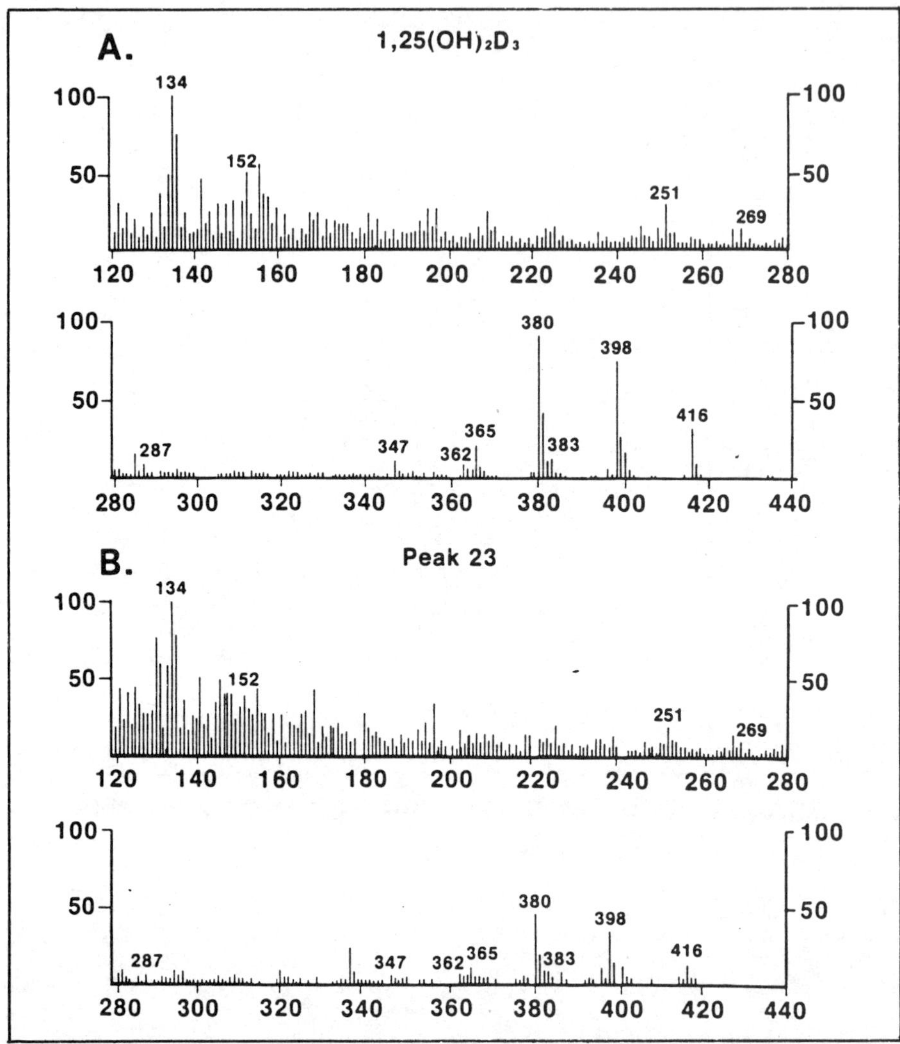

Figure 11. Mass spectrum determinations of authentic $1,25(OH)_2D_3$ (panel A) and purified peak 23 (panel B). Fragmentations characteristic of $1,25(OH)_2D_3$ are indicated and described in the text.

b. Methods. To determine the effect of chronic $1,25(OH)_2D_3$ administration on the *in vitro* metabolism of $25(OH)D_3$, $24,25(OH)_2D_3$, and $1,25(OH)_2D_3$ by liver and kidney, we infused animals with either vehicle alone or $1,25(OH)_2D_3$ at the rate of 75 pmol/d for 12 days. At the end of this time the livers and kidneys were perfused *in vivo* with 10 ml ice-cold homogenizing buffer, excised, passed through a Krebs tissue press, and homogenized using a teflon pestle in 2 vol of the same buffer: 100 mM KCl, 10 mM Tris-HCl, 10 mM NaH_2PO_4, 10 mM Na-pyruvate, 4 mM $MgSO_4$ and 10 mM EDTA with $CaCl_2$ added to a free calcium concentration of 0.01 mM at pH 6.8. 1 ml aliquots of the homogenized tissue were combined with

0.1 μCi of [23,24] [^{3}H]25(OH)D_3 (85Ci/mmol), [26,27] [^{3}H]25(OH)D_3 (158Ci/mmol), [23,24] [^{3}H] 24,25$(OH)_2D_3$ (64Ci/mmol), [26,27] [^{3}H]1,25$(OH)_2D_3$ (158Ci/mmol), or [23,24] [^{3}H]1,25$(OH)_2D_3$ (110Ci/mmol) in 10 μl of ethanol, gassed with oxygen, and incubated (in duplicate) in a shaking water bath at 37°C for varying periods of time. The incubations were stopped by adding 3 ml of a 2:1 mixture of methanol:chloroform, the samples shaken vigorously and phase separation accomplished by adding an additional 1 ml of chloroform and 0.8 ml H_2O. After centrifugation at 3,000 rpm for 5 minutes in a clinical bench top centrifuge, the lower (chloroform) phase was removed and the upper aqueous phase reextracted twice with 1 ml chloroform. The chloroform extracts were combined, dried under nitrogen and resolubilized in 150 μl hexane/isopropanol (90:10) in preparation for HPLC. The chromatographic system consisted of a Zorbax Sil column (Du Pont Instruments, Wilmington, DE) and a solvent system of hexane/isopropanol run in a concave gradient mode (setting 9, Waters 660 programmer, Waters Associates, Milford, MA) from 97:3 to 90:10 at a flow rate of 2 ml/min. Tritium in the column effluent was monitored continuously using a FLO-ONE radioactive flow counter (Radiomatic Instrument and Chemical Co., Tampa, FL).

c. Results. Chronic 1,25$(OH)_2D_3$ administration (12 days, 75 pmol/d) resulted in virtually no change in the *in vitro* metabolism of 25(OH)D_3, 24,25$(OH)_2D_3$, and 1,25$(OH)_2D_3$ by liver homogenates (Table 5). However, 1,25$(OH)_2D_3$ administration substantially increased the *in vitro* metabolic conversion of 25(OH)D_3 to 24,25$(OH)_2D_3$ in renal homogenates. In preparations of renal tissue from animals infused with vehicle alone 89.4% of the original 25(OH)D_3 added to the incubation media was recovered intact, while 3.1% was converted to 24,25$(OH)_2D_3$. In preparations of renal tissue from animals chronically infused with 1,25$(OH)_2D_3$ for 12 days at a rate of 75 pmol/d, only 71.7% of the original 25(OH)D_3 added to the incubation media was recovered intact while 21.6% was converted to 24,25$(OH)_2D_3$. Catabolism of 24,25$(OH)_2D_3$ in renal preparations in two of three independent experiments was not significantly different between vehicle and 1,25$(OH)_2D_3$ infused animals although an increase in catabolism of 24,25$(OH)_2D_3$ by the 1,25$(OH)_2D_3$ infused kidney homogenates was observed in a third experiment. Catabolism of 1,25$(OH)_2D_3$, which was rapid in control animals, was further increased in animals chronically infused with 1,25$(OH)_2D_3$. Of interest is that 1,25$(OH)_2D_3$ is metabolized to similar, if not identical, metabolites by renal homogenates and keratinocytes.

We then compared the metabolism of 25(OH)D_3 and 1,25$(OH)_2D_3$ by renal homogenates using substrates which differed in the location of the tritium label. Our purpose was to determine whether tritium in the C23 and C24 positions would reduce hydroxylation at C24 compared to tritium located in the C26,C27 positions. The experiment was comparable to that described above except that both [26,27] [^{3}H] and [23,24] [^{3}H] 25(OH)D_3 and 1,25$(OH)_2D_3$ were employed. Incubations with the labeled 25(OH)D_3 were 2 hours as before but were reduced to 30 minutes with the 1,25$(OH)_2D_3$. The

Table 5. In vitro metabolism of $25(OH)D_3$, $24,25(OH)_2D_3$ and $1,24,25(OH)_2D_3$ by liver and kidney homogenates from rats chronically infused with $1,25(OH)_2D_3$ or vehicle for 12 days.

			Percent substrate recovered				
Tissue	Substrate	$1,25(OH)_2D_3$ dose pmol/d	$25(OH)D_3$	$24,25(OH)_2D_3$	$1,25(OH)_2D_3$	Other	Aqueous
LIVER	$25(OH)D_3$	0	89.1	0	0	5.0	2.0
		75	91.1	0	0	3.7	1.5
	$24,25(OH)_2D_3$	0	0	67.3	0	22.7	4.6
		75	0	69.3	0	26.5	2.2
	$1,25(OH)_2D_3$	0	0	0	96.9	1.6	2.4
		75	0	0	97.8	0.8	2.0
KIDNEY	$25(OH)D_3$	0	89.4	3.1	0.2	0	2.9
		75	71.7	21.6	0.2	0	4.1
	$24,25(OH)_2D_3$	0	0	75.7	0	13.0	9.0
		75	0	80.2	0	14.2	11.3
	$1,25(OH)_2D_3$	0	0	0	6.0	5.5	62.5
		75	0	0	0	0	86.6

Values (mean of duplicate determinations) are expressed as a percentage of the total counts incubated with the homogenates. In this experiment the incubations were 2 h at 37°C. $25(OH)D_3$ and $1,25(OH)_2D_3$ were radiolabeled in C26, C27 whereas $24,25(OH)_2D_3$ was labeled in C23, C24.

results show that in all cases the 23,24 labeled substrate, be it $25(OH)D_3$ or $1,25(OH)_2D_3$, was metabolized more slowly than the 26,27 labeled substrate. The major metabolite of $25(OH)D_3$ produced by renal homogenates, especially after $1,25(OH)_2D_3$ infusion, was $24,25(OH)_2D_3$. The major metabolites of $1,25(OH)_2D_3$ produced by renal homogenates were peak 28 (thought to be either 24-keto,$1,23,25(OH)_3D_3$ or $1,23,25(OH)_3D_3$ and $1,24,25(OH)_3D_3$. In the formation of these metabolites the hydrogen (or tritium) at C23 or C24 is removed. The substitution of tritium for hydrogen appears to retard this process.

Because of the apparent increase in $1,25(OH)_2D_3$ metabolism by the renal homogenates after $1,25(OH)_2D_3$ infusion, we elected to study this process more closely. Rats were infused with $1,25(OH)_2D_3$ (75 pmol/day) or vehicle alone for 12 days as before (Table 6). The kidneys were obtained and homogenized as before. One ml aliquots of the homogenate were placed into 15 ml tubes. Following a 5 minute preincubation at 37°C, 0.1μCi of 26,27(n)-[^{3}H]$1,25(OH)_2D_3$ (sp act 158 Ci/mmol) was added to start the reaction. After 0, 5, 10, 15, 30 and 60 minutes, the reaction was stopped by addition of 3 ml of a 2:1 methanol:chloroform mixture. Extraction and HPLC were preformed as before. The results depicted in Figure 12 demonstrate that $1,25(OH)_2D_3$ infusion enhances its own metabolism by the kidney. Even in the control rat kidneys, the metabolism of the radiolabeled [^{3}H]$1,25(OH)_2D_3$

Table 6.

A. Metabolism of radiolabeled $25(OH)_2D_3$ (% Recovered)

	$25(OH)D_3$	$24,25(OH)_2D_3$
Vehicle infused		
^{3}H-26,27	89.2 ± .6	.5 ± .2
^{3}H-23,24	95.5 ± .1	.05 ± .05
$1,25(OH)_2D_3$ infused		
^{3}H-26,27	80.6 ± .3	2.9 ± .1
^{3}H-23,24	92.0 ± .6	1.2 ± .3

B. Metabolism of radiolabeled $1,25(OH)_2D_3$ (% Recovered)

	$1,25(OH)_2D_3$	Peak 28	$1,24,25(OH)_3D_3$
Vehicle infused			
^{3}H-26,27	74.9 ± .3	12.5 ± .1	7.7 ± .2
^{3}H-23,24	93.1 ± .7	0.2 ± .0	3.8 ± .5
$1,25(OH)_2D_3$ infused			
^{3}H-26,27	47.5 ± 2.1	25.0 ± 2.8	13.4 ± .2
^{3}H-23,24	58.8 ± 3.2	7.0 ± 2.0	10.8 ± .5

Rats were infused with $1,25(OH)_2D_3$ (75 pmol/day) or vehicle for 12 days before the kidneys were removed and homogenized. The incubation with $25(OH)D_3$ lasted 2 h; the incubation with $1,25(OH)_2D_3$ lasted 30 min.

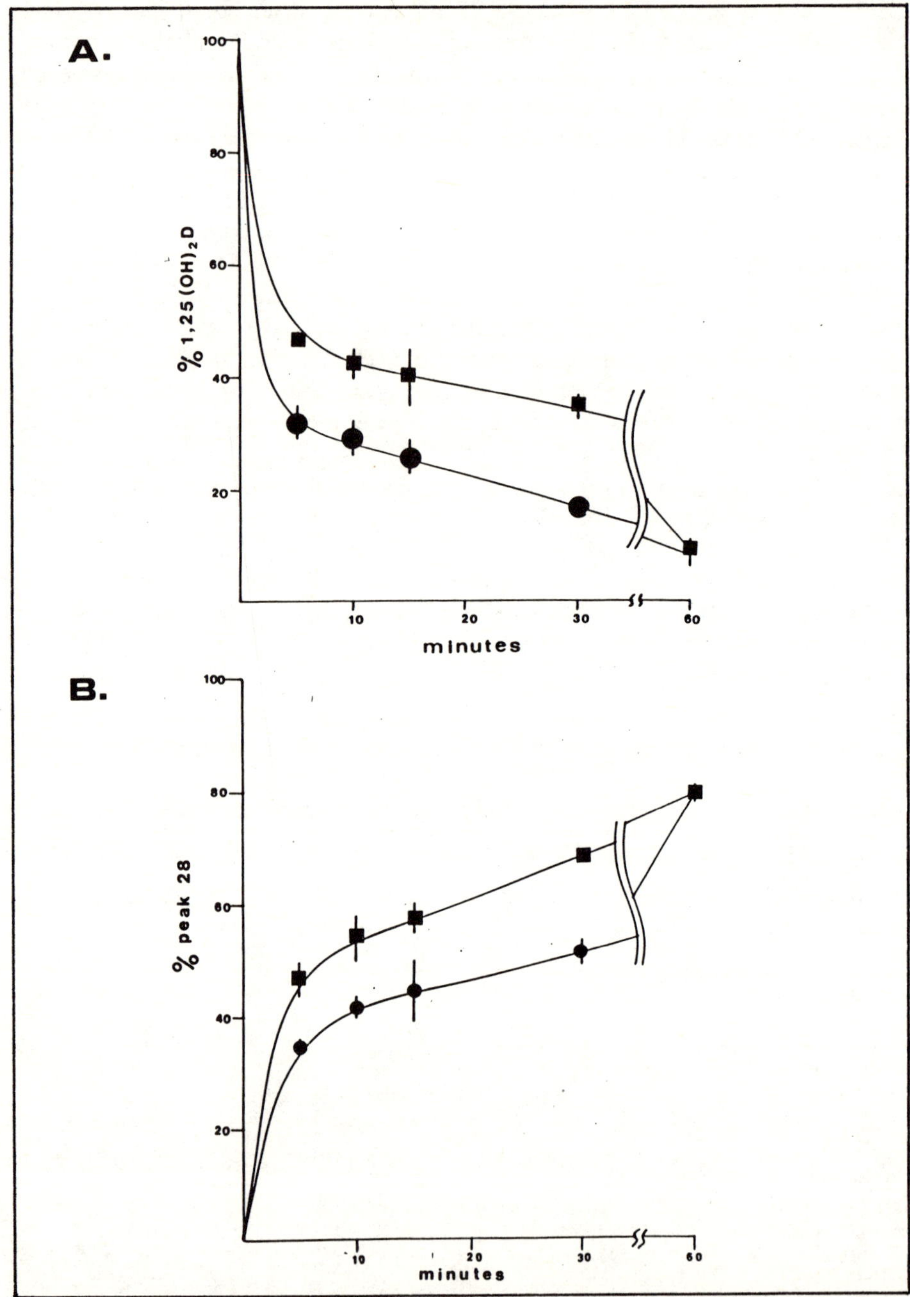

Figure 12. Enhancement by $1,25(OH)_2D_3$ of its own metabolism by rat kidney homogenates. Homogenates from rat kidneys infused with 75 pmol $1,25(OH)_2D_3$ for 12 days (●) or control kidneys (■) were incubated with [^{3}H] $1,25(OH)_2D_3$ for various amounts of time. A. Catabolism of $1,25(OH)_2D_3$ as a function of time. Control (■) [^{3}H] $1,25(OH)_2$ D incubated (●). B. Appearance of "peak 28" as a function of time. Control (●) [^{3}H] $1,25(OH)_2$ D incubated (■).

is rapid, reaching 50% disappearance by 5 minutes in this experiment. In the $1,25(OH)_2D_3$ infused rat kidneys, the recoverable radiolabeled $[^3H]1,25(OH)_2D_3$ is only 33% after 5 minutes. After 5 minutes the rate of metabolism appears to slow, such that by 30 minutes 35% and 17% of the added $[^3H]1,25(OH)_2D_3$ can be recovered from the control and $1,25(OH)_2D_3$ infused rat kidney homogenates.

The $[^3H]$ $1,25(OH)_2D_3$ is converted primarily to peak 28. As shown in the figure, peak 28 increases rapidly in the first 5 minutes of incubation, and this increase is more marked in the kidneys from rats infused with $1,25(OH)_2D_3$.

Screening of cell lines for vitamin D metabolism

Several renal carcinoma cell lines (1072F, 769P, 786O, CAKI) were compared to nontransformed renal cell lines (NTK or OK cells) for the metabolism of $[^3H]$ $25(OH)D_3$ in an effort to find a model of renal vitamin D metabolism which could readily be studied *in vitro*. 1072F, 769P, 786O, and CAKI were derived from renal cell carcinomas; NTK was derived from normal appearing tissue adjacent to a human renal cell carcinoma, and the OK cell line was derived from opossum kidney.

These cells were grown to confluence in 35 mm dishes using Dulbecco's MEM containing 5% fetal calf serum. They were then switched to serum free medium 48 hours prior to assay. On the day of assay, the cells were placed in 0.5 ml fresh serum free medium to which 0.1 μCi of $[^3H]$ $25(OH)D_3$ in 5 μl ethanol was added. They were incubated for 4 hours at 37°C in 5% CO_2: 95% air. Cells and media in each dish were extracted and the chloroform phase was analyzed for $[^3H]$ vitamin D_3 metabolites by HPLC. HPLC was performed using a Du Pont Zorbax Sil column and a concave gradient of 97:3 to 90:10 hexane:isopropanol at 2 ml/min flow rate. Eluted radioactivity was quantitated using a Radiomatic FLO-ONE detector.

The results are depicted in Table 7. All cell lines produced a metabolite which coeluted with $24,25(OH)_2D_3$. None of these cell lines produced $1,25(OH)_2D_3$. Several of these cell lines produced a metabolite with the chromatographic behavior of $25,26(OH)_2D_3$.

Table 7. $25(OH)D_3$ metabolism by renal cell lines.

	% Recovered			
Human	$25(OH)D_3$	$24,25(OH)_2D_3$	$25,26(OH)_2D_3$	Peak 28
1072 F	93.2	4.5	0.1	—
769 P	98.0	1.6	—	—
786 O	94.9	4.5	—	—
CAKI	76.8	8.6	1.2	4.1
NTK	94.4	3.8	0.2	—
Opossum				
OK*	53.7	9.4	1.6	2.3

*In this experiment the OK cells were incubated with 10^{-10} M $1,25(OH)_2D_3$ for 16 h prior to assay to increase the metabolism of $25(OH)D_3$.

REFERENCES

Atuma, S.S., Lundstrom, K. and Lindquist, J. (1979). Cited in: Vanhaelen-Fastre and M. Van Haelen (1980). Separation and determination of the D vitamins by HPLC. In: *Steroid Analysis by HPLC - Recent applications*. M.P. Kautsky, (ed). Marcel Dekker, Inc., New York, Bard, pp. 173-251.

Bikle, D.D., Nemanic, M.K., Gee, E. and Elias, P. (1986a). 1,25-dihydroxy vitamin D_3 production by human keratinocytes. *J. Clin. Invest.* 78, 557-566.

Bikle, D.D., Nemanic , M.K., Whitney, J.A. and Elias, P.M. (1986b). Neonatal human foreskin keratinocytes produce 1,25-dihydroxy vitamin D_3. *Biochemistry* 25, 1545-1548.

Bligh, E.G. and Dyer, W.G. (1959). A rapid method of total lipid extraction and purification. *Can. J. Biochem. Physiol.* 37, 911-917.

Bollinger, H.F. (1965). In: *Thin layer chromatography*. Stahl E.,(ed). Academic Press, New York. pp. 223.

Clemens, T.L., Adams, J.S., Horiuchi, N., Gilchrest, B.A., Cho.,H., Tsuchiya, Y., Matsuon, Suda, T. and Holick, M.F. (1983). Interaction of 1,25 dihydroxy vitamin D_3 with keratinocytes and fibroblasts from skin of normal subjects and subjects with vitamin D dependent rickets Type II. A model for study of mode of action of 1,25 dihydroxy vitamin D. *J. Clin. Endocrinol. Metab.* 56, 824-830.

DeLuca, H.F. and Blunt, J.W. (1971). Vitamin D. In: *Methods in Enzymology. XVIII Part C.*, Mc Cormick and Wright (eds). S.P. Colowick and N.O. Kaplan; (Pub). Academic Press, New York, London, pp. 709-733.

DeLuca (1983). In: *Bone and Mineral Research,* W.A. Peck, (ed). *Amsterdam Excerpta Medica.* pp. 7-73.

Egaas, E. and Lambertsen, G. (1979). Cited in: Vanhaelen-Fastre and M. Van Haelen, (1980). Separation and determination of the D vitamins by HPLC. In: *Steroid Analysis by HPLC - Recent applications*. M.P. Kautsky, (ed). Marcel Dekker, Inc., New York, Bard, pp. 173-251.

Frankel, T.L., Mason, R.S., Hersey, P., Murray, E. and Posen, S. (1983). The synthesis of vitamin D metabolites by human melanoma cells. *J. Clin. Endocrinol. Metab.* 57, 627-31.

Fraser, D.R. and Kodicek, E. (1970). Unique biosynthesis by kidney of a biologically active vitamin D metabolite. *Nature* (London) 228, 764-766.

Gray, T.K., Lester, G.E. and Lorenc, R.S. (1979). Evidence for extrarenal 1a hydroxylation of 25 hydroxy vitamin D in pregnancy. *Science* (Washington, D.C.) 204, 1311-13.

Hashmi, Manzhur-Ul-Haque (1972). *Assay of vitamins in pharmaceutical preparations*. John Wiley and Sons, New York, pp. 324-359

Holick, M.F., MacLaughlin, J.A., Clark, M.B., Holick, S.A., Potts, J.R.,Jr. and Elias, P.M. (1980). Photosynthesis of previtamin D_3 in human skin and the physiologic consequences. *Science* 210, 203-205.

Holick, M.F., Richard, N.M., NcNeill, S.C., Holick, S.A., Frommer, J.E., Henley, J.W. and Potts, J.R., Jr. (1979). Isolation and identification of previtamin D_3 from the skin of rats exposed to UV radiation. *Biochemistry*. 18, 1003-1008.

Holick, M.F. (1971). The cutaneous photosythesis of previtamin D_3: a unique photoendocrine system. *J. Invest. Dermatol.* 77, 51-58.

Horst, R.L. (1983). Resolution and quantitation of vitamin D and vitamin D metabolites. In: *Assay of calcium-regulating hormones,* D.D. Bikle, (ed). Springer-Verlag, New York, Berlin, Tokyo. pp. 21-47.

Hosomi, J., Horch, J., Able, Suda, T., Kurolu, T. (1983). Regulation of terminal differentiation of cultured mouse epidermal cells by 1,25 dihydroxy vitamin D_3. *Endocrinology* 113, 1950-1957.

Howard, G.A., Turner, R.T., Sherrard, D.J. and Baylink, D.J. (1981). Human bone cells in culture metabolize 25(OH)D to $1,25(OH)_2D$ and $24,25(OH)_2D$. *J. Biol. Chem.* 256, 7738-40.

Kulkowski, J.A., Chan, T., Martinez, J., Ghazanian, J.G. (1979). Modulation of 25(OH)D-24 hydroxylase by aminophylline: A cytochrome P 450 monooxygenase system. *Biochem. Biophys. Res. Commun.* 90, 50-57.

Lambert, P.W., Stern, P.H., Avioli, R.C., Brackett, N.C., Turner, R.T., Greene, A., Tiu, I.Y. and Bell, N.H. (1982). Evidence for extrarenal production of 1-a $25(OH)_2D$ in man. *J. Clin. Invest.* 69, 722-25.
Littledike, E.T. and Horst, R.L. (1982). Metabolism of vitamin D in nephrectomized pigs given pharmacological amounts of vitamin D. *Endocrinology* (Baltimore) 111, 2008-13.
Mason, R.S., Frankel, T., Chan, Y.-L., Lissner, D. and Posen, S. (1984). Vitamin D conversion by sarcoid lymph node homogenate. *Ann. Intern. Med.* 100, 59-61.
Mc Lauglin, J.A., Anderson, R.R., Holick, M.F. (1982). Spectral character of sunlight modulates photosynthesis of previtamin D_3 and its photoisomers in human skin. *Science* 216, 1001-1004.
McCollum, E.V., Simmonds, N., Shipley, P.G. and Park, E.A. (1922). A delicate biological test for calcium depositing substances. *J. Biol. Chem.* 54, 41-48.
Mellanby, E. (1918). The part played by an ''accessory factor'' in the production of experimental rickets. *J. Physiol.* 52, 11-14.
Menicagli, C., Silverstri, S. and Nucci, T. (1976). Cited in: Vanhaelen-Fastre and M. Van Haelen (1980). Separation and determination of the D vitamins by HPLC. In: *Steroid Analysis by HPLC - Recent applications.* M.P. Kautsky, (ed). Marcel Dekker, Inc., New York, Bard, pp. 173-251, 252.
Napoli, J.L., Pramanik, B.C., Royal, P.M., Reinhardt, T.A., Horst, R.L. (1983). Intestinal synthesis of 24-keto-1,25 dihydroxyvitamin D_3. *J. Biol. Chem.* 258: 9100-9107
Nemanic, M.K., Whitney, J., and Elias, P.M. (1985). *In vitro* synthesis of vitamin D_3 by cultured human keratinocytes and fibroblasts. Action spectrum and effect of AY-9944. *Biochem. Biophys. Acta* 841, 267-277.
Reeve, S., Tanaka, Y. and DeLuca, H.F. (1983). Studies on the sites of $1,25(OH)_2D$ synthesis *in vivo*. *J. Biol. Chem.* 258, 3615-17.
Rheinwald, J. and Green, H. (1975). Serial cultivation of human epidermal keratinocytes: The formation of keratinizing colonies from single cells. *Cell* 6 331-343.
Sheppard, A.J. and Hubbard, W.D. (1971). Gas chromatography of vitamin D_2 and D_3. In: *Methods in Enzymology*. XVIII Part C., Mc Cormick and Wright, (eds). S.P. Colowick and N.O. Kaplan, (Pub). Academic Press, New York, London, pp. 733-738.
Smith, E.L., Walworth, N.C., Holick, M.F. (1986). Effect of 1,25 dihydroxy vitamin D_3 on the morphologic and biochemical differentiation of cultured human epidermal keratinocytes grown in serum free conditions. *J. Invest. Dermatol.* 86, 709-714.
Strong, J.S. (1976). Combined assays for vitamin A, D and E in multivitamin preparation with separation in reverse phase partition chromatography. *J. Pharm. Sci.* 65, 968-974.
Stumpf, W.E., Sar, M., Reid, R.A., Tanaka, Y., and DeLuca, H.F. (1979). Target cells for 1,25 dihydroxy vitamin D in the intestinal tract, stomach, kidney, skin, pituitary, and parathyroid. *Science* 206, 1188-1190.
Vanhaelen-Fastre and Van Haelen, M. (1980). Separation and determination of the D vitamins by HPLC. In: *Steroid Analysis by HPLC - Recent applications.* M.P. Kautsky, (ed). Marcel Dekker, Inc., New York, Bard, pp. 173-251.
Windaus, A. and Bock, F. (1937). Uber das provitamin aus dem slerin der schweines schwarte. Hoppe-Seyler's Z. *Physiol Chem.* 245, 168-172.

Progress in HPLC, Vol. 3, pp. 129-147
Parvez et al. (Eds)

High-performance liquid chromatography of radiolabeled nicotine enantiomers and their metabolism in the guinea pig to N-methylated products

PETER A. CROOKS and KENNETH C. CUNDY
Division of Medicinal Chemistry and Pharmacognosy,
College of Pharmacy, University of Kentucky,
Lexington, KY, USA

INTRODUCTION

The emergence of flow through radioactive detectors has resulted in the development of rapid and sensitive high-performance liquid chromatographic methods for the identification and quantitation of radioactive products in metabolic tracer studies. In addition, the use of an electronic stream splitter, which diverts a certain percentage of the eluate from the HPLC column to collection vials, affords an efficient means of purifying and isolating radiolabeled components in a mixture.

We have recently utilized flow through radioactivity detection in studying a chirodiastaltic phenomenon in the chromatographic analysis and separation of radiolabeled nicotine enantiomers in a totally achiral system. This has led to a new methodology for the separation and isolation of radiochemically and optically pure samples of nicotine enantiomers. The application of this phenomenon in studying the biotransformation of nicotine enantiomers has also afforded a unique analytical tool for detecting the stereospecific metabolism of nicotine to quaternary N-methylated metabolites, a metabolic pathway that has been little studied until recently.

THE CHIRODIASTALTIC PHENOMENON

Despite the widely held belief that optical isomers are chemically identical, the unusual phenomenon observed in this study is not entirely without precedent. A number of investigators (Horeau and Guette, 1974; Kabachnik et al., 1976; Williams et al., 1969) have reported a similar phenomenon in the NMR spectroscopy of racemates and pure enantiomers of certain optically active compounds. Whenever there is an excess of one of the enantiomers in a sample, the NMR spectrum splits; smaller signals appear at different chemical shifts for the lesser of the two isomers present, with a relative intensity proportional to its abundance in the mixture. As the sample approaches racemic composition (i.e., equal amounts of both (+)- and (−)-isomers), the signals of each enantiomer become equal in intensity and superimpose. The explanation of this result can best be understood by the way in which two different optical isomers interact with their environment. One enantiomer is in contact with molecules that are predominantly identical to itself. These molecules would be in a highly ordered state, whether directly interacting with each other (as in a pure liquid) or surrounded by solvent molecules (as in a solution). In either case, the enantiomer in question would fit easily into such an arrangement. The other enantiomer however, is in contact with molecules of the opposite configuration. Integration into the structure of the surrounding liquid or solution would be energetically unfavorable and would result in a loss of order in its environment. The result is that the average magnetic field around each of the enantiomers is very different, giving rise to different NMR spectra.

The above explanation is also applicable to a wide range of physicochemical properties which appear to be different for mixed enantiomers, including solubility, melting point, absorption, etc. (Horeau and Guette, 1974). In all cases, there is a differentiation in the way in which the two optical isomers interact with their environment; such interaction have been termed chirodiastaltic. All chirodiastaltic phenomena require an imbalance in the total optical activity of the system, whether it be due to the presence of another optically active moiety (e.g., an enzyme molecule, a chiral resolving agent, etc.) or merely to the presence of an excess of one enantiomer.

SEPARATION OF RADIOLABELED NICOTINE ENANTIOMERS BY ACHIRAL HIGH-PERFORMANCE LIQUID CHROMATOGRAPHY

Nicotine (1) is the principal tobacco alkaloid, which exists naturally in the plant as the 2′-S-(−)-enantiomer (1a); consequently, the S-(−)- enantiomer is the only commercially available form of nicotine. As a result, until quite recently, the majority of metabolic studies involving nicotine have utilized only the S-(−)-isomer. Interestingly, those few studies that have also involved the use of the R-(+)-enantiomer (1b) have indicated a marked stereoselectivity in certain metabolic pathways (Gorrod and Jenner, 1975), and recent results from our own laboratories (Cundy et al., 1984, 1985, a,b) have established both stereoselective and stereospecific differences in the metabolism of nicotine enantiomers in a number of animal species. These observations take on more significance with regard to the metabolism and toxicology of 'smoke nicotine', which has been shown to contain small amounts of R-(+)-nicotine formed from S-(−)- nicotine, presumably via pyrolytic racemization during the burning of the cigarette (Klus and Kuhn, 1977).

At the onset of our studies, our initial aim was to develop suitable synthetic routes to radiolabeled R-(+)- and S-(−)-nicotine enantiomers, and to

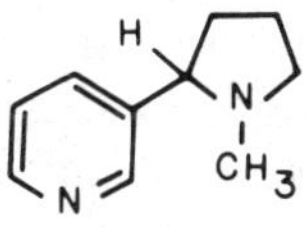

S-(-)-Nicotine
1a

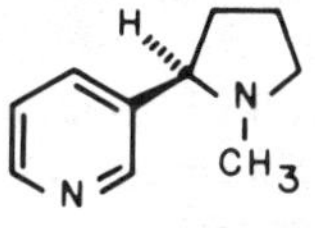

R-(+)-Nicotine
1b

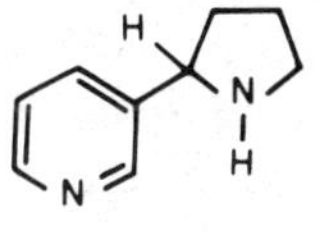

Nornicotine
2

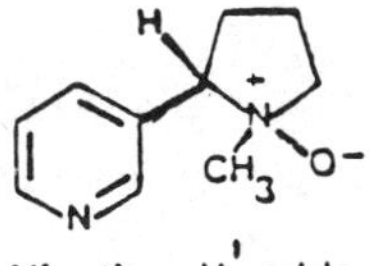

Nicotine-N'-oxide
4

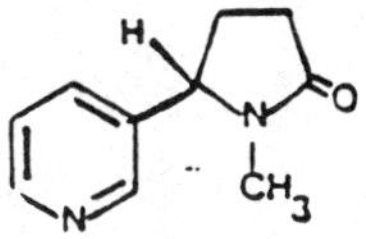

Cotinine
3

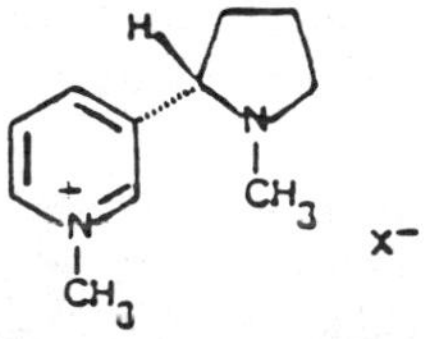

S-N-methylnicotinium ion
5a

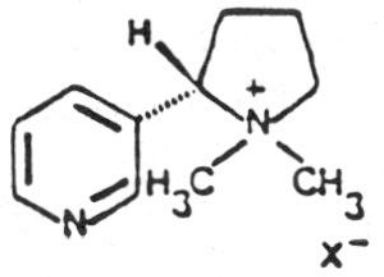

S-N'-methylnicotinium ion

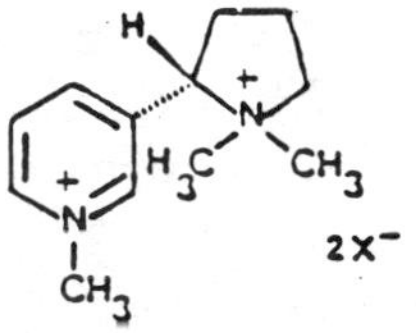

S-N,N'-dimethylnicotinium ion

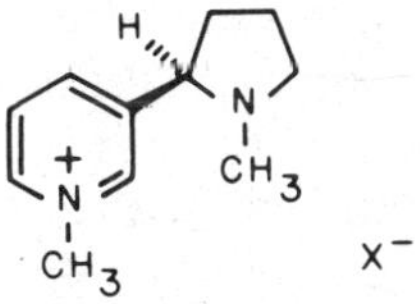

R-N-methylnicotinium ion
5b

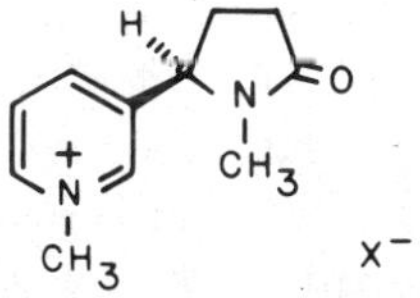

S-N-methylcotininium ion

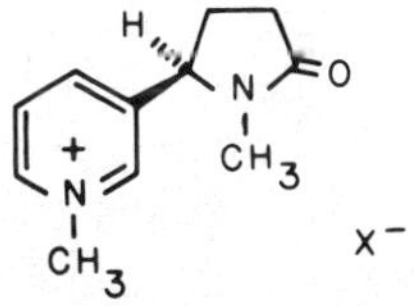

R-N-methylcotininium ion
7

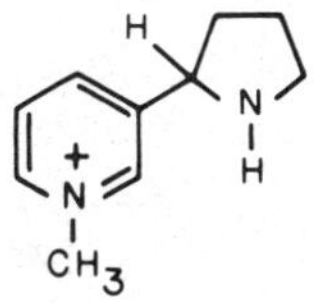

N-methylnornicotinium ion
6

utilize these products in *in vitro* and *in vivo* experiments that were designed to compare and contrast the metabolic profile of each nicotine enantiomer. At the time this work was initiated, no tritiated forms of R-(+)-nicotine were commercially available, although ^{3}H-(−)-nicotine could be otained from New England Nuclear and Amersham. Also, no optically pure forms of ^{14}C-nicotine are yet available, since all direct syntheses of this compound reported to date have involved the formation of an intermediate which affords a racemic compound.

A pure ^{14}C-labeled enantiomer of nicotine could theoretically be prepared from optically pure nornicotine (2). However, naturally occurring nornicotine is not optically pure, and resolution of its enantiomers has been achieved only via the synthesis of a precursor that is more susceptible to fractional crystallization (Jacob, 1982). Several studies appeared in the literature in the 1960's which claim to have used [^{14}C-NCH_3]-(−)-nicotine (Hansson and Schmitterlow, 1962; Schmitterlow and Hansson, 1962; McKennis et al., 1963; Decker and Sammeck, 1964; Appelgren et al., 1962; Anderson et al., 1965). However, consideration of the source of this material (McKennis et al, 1961) shows that it was obtained from methylation of natural nornicotine. Thus all this work was presumbaly performed on a partially racemic radiolabeled material.

We have recently observed a chirodiastaltic phenomenon in the HPLC analysis of racemic ^{14}C-labeled nicotine, which has enabled us to prepare optically pure radiolabeled nicotines (Cundy and Crooks, 1983). When racemic radiolabeled nicotine is coinjected onto a number of different achiral HPLC systems together with an unlabeled *S*-(−)-nicotine standard, the radioactivity eluting from the column is divided into two distinct peaks. One of these peaks coelutes with the unlabeled standard, and one elutes considerably later (Fig. 1). This phenomenon appears to be due to a differential enantiomeric association between like and unlike optical isomers of nicotine, in an enantiomeric mixture where one isomer is in excess. The practical application of this phenomenon has been utilized in the direct resolution of ^{14}C-labeled racemic nicotine into its enantiomers, using the preparative HPLC analytical system illustrated in Figure 2.

As can be seen in Figure 3, the radioactivity eluting from the preparative column after injection of ^{14}C-(±)-nicotine with an authentic standard of S-(−)-nicotine, was divided into two distinct peaks, one coeluting with unlabeled *S*-(−)-nicotine standard, and one eluting considerably later. Fractions corresponding to this second radioactive peak were identified, and the contents of the appropriate scintillation vials were combined. The resulting solution was concentrated under reduced pressure to remove methanol and adjusted to ~ pH 9 with 1 N NaOH (this treatment does not affect the chiral integrity of the nicotine molecule). The alkaline solution was then applied to a column of Amberlite XAD-2 resin and the column was rinsed with water to remove any salt originating from the HPLC mobile phase (this treatment does not elute the ^{14}C-labeled nicotine). The radiolabeled nicotine was then eluted from the column using methanol; aliquot parts of fractions were

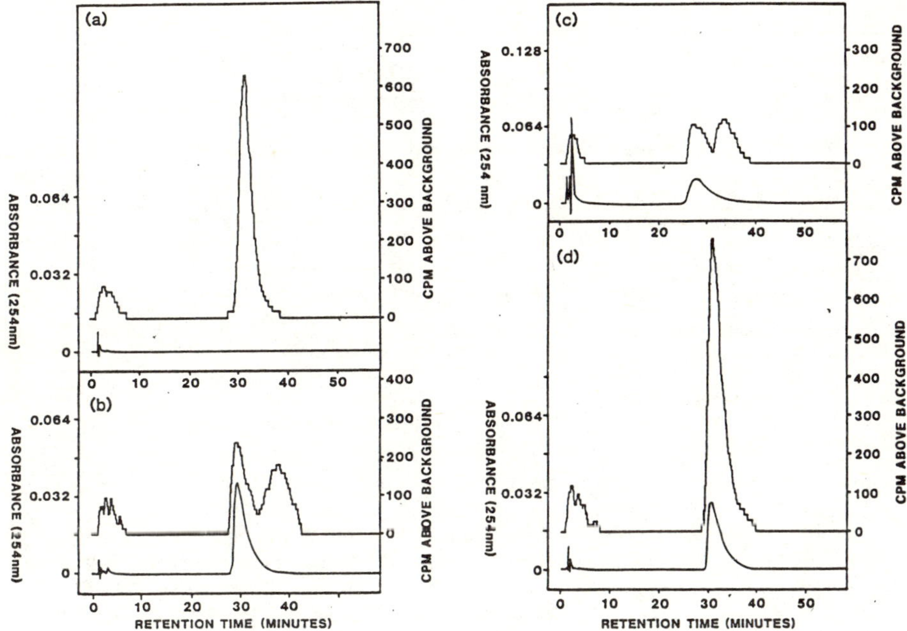

Figure 1. Radiochromatograms obtained on Partisil-PXS cation-exchange column for [^{14}C-NCH_3] -(±)-nicotine-(+)-bitartrate (15 μl, 1 μCi/ml): a) alone; b) with unlabeled (−)-nicotine (5 μl, 0.4 mg/ml); d) with unlabeled racemic nicotine (5 μl, 0.4 mg/ml). Eluent: 0.06M sodium acetate/methanol (70:30), pH 6.8, 2.0 ml/min. Detection: ^{14}C by Radiomatic FLO-ONE Model HS; UV at 254 nm.

analyzed. A sample of the *R*-(+)- [^{14}C-NCH_3]-nicotine isolated from XAD-2 chromatography was examined by SCX HPLC analysis. As can be seen (Fig. 4), the ^{14}C activity was almost completely confined to a single peak, eluting after the coinjected unlabeled *S*-(−)-nicotine standard. A very small proportion of the ^{14}C activity coeluted with the unlabeled S-(−)-nicotine. Integration of the radioactive peaks allowed assessment of the radiochemical and radiooptical purity of the isolated material (Table 1).

An identical procedure to that described above was applied to the preparative isolation of the *S*-(−)-enantiomer of [^{14}C-NCH_3]-nicotine. However, in this case, unlabeled *R*-(+)-nicotine was coinjected with the racemic radiolabeled material onto the SCX HPLC system. The ^{14}C activity eluting from the column was again divided into two distinct peaks, of which the second peak corresponded to *S*-(−)-[^{14}C-NCH_3]-nicotine. Thus fractions corresponding to this radioactive peak were identified and analyzed as before. Once again, HPLC analysis showed that the ^{14}C activity was almost completely confined to a single peak, eluting after a coinjected unlabeled R-(+)-nicotine standard, and integration of the radioactive peaks permitted calculation of the radiochemical and radiooptical purity of the product (Table 1). The above study illustrates that racemic radiolabeled nicotine can be resolved into its two enantiomers on a totally achiral HPLC system, when

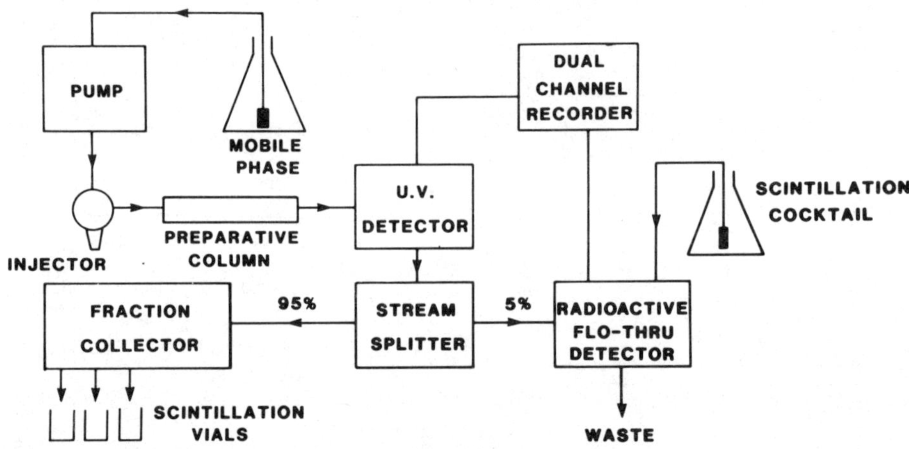

Figure 2. Schematic diagram of the HPLC system. Separations were carried out on a Magnum 9 SCX 10/50 preparative cation-exchange column (Whatman). The mobile phase was isocratic 0.3M sodium acetate/methanol (70:30) adjusted to pH 6.8 with glacial acetic acid at a flow rate of 2.0 ml/min. Samples were introduced via a Rheodyne loop injector (Rheodyne, Cotati, CA) equipped with a 50 μl loop. A Model ES stream splitter (Radiomatic, Tampa, FL) was used to divert 5% of the column effluent emerging from the UV detector to a FLO-ONE Model HS radioactive flow-through detector (Radiomatic). Here, it was mixed, in a 1:4 ratio by volume with Scintiverse LC scintillation cocktail (Fisher) and ^{14}C content was monitored continuously. The output of the UV detector (254 nm) and the flow-through radioactivity detector were recorded simultaneously. The remaining 95% of column effluent emerging from the UV detector was diverted to a fraction collector.

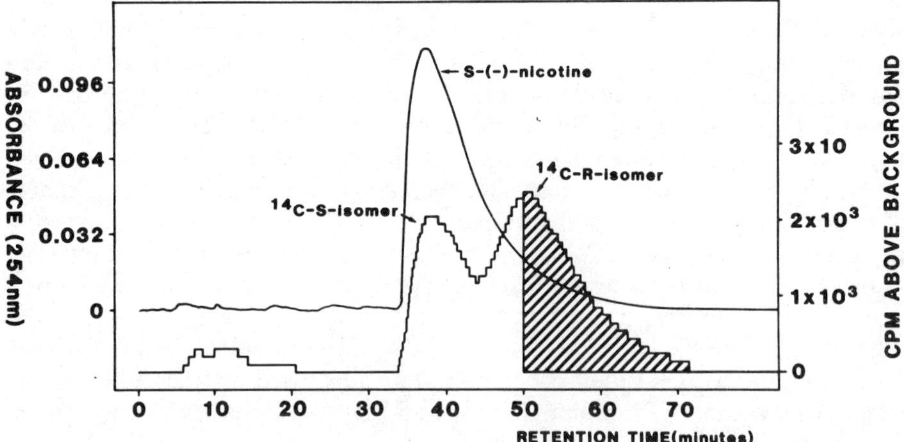

Figure 3. Preparative Partisil-10 SCX cation-exchange high-pressure liquid chromatographic separation of R-(+)- and S-(−)- [^{14}C-NCH_3]-nicotines, after coinjection with authentic unlabeled S-(−)-nicotine standard. Shaded area indicates fraction isolated.

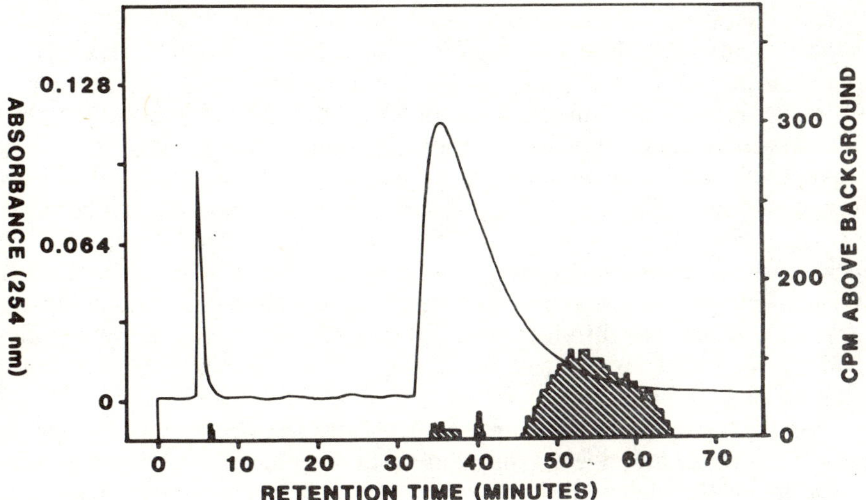

Figure 4. Partisil 10 SCX cation-exchange high-pressure liquid radiochromatographic analysis of resolved R-(+)-[^{14}C-NCH_3]-nicotine after coinjection with authentic unlabeled S-(−) nicotine standard. Shaded area indicates radioactivity in fractions isolated.

coinjected with one pure enantiomer of unlabeled nicotine. Thus, by coinjecting racemic radiolabeled nicotine with unlabeled S-(−)-nicotine, the second radioactive peak eluting from the column corresponds to the R-(+)-enantiomer of radiolabeled nicotine. This peak can be isolated and is uncontaminated by the other radiolabeled enantiomer. The radioacive peak coeluting with unlabeled S-(−)-nicotine corresponds to the *S*-(−)-enantiomer of the radiolabeled nicotine. This peak is also uncontaminated by the other radiolabeled enantiomer,but has a greatly reduced specific activity due to its dilution by the unlabeled standard. Alternatively, by coinjecting racemic radiolabeled material with unlabeled R-(+)-nicotine standard, the second radioactive peak eluting corresponds to the pure S-(−)-enantiomer of the radiolabeled nicotine.

The resolution of racemic radiolabeled nicotine on an otherwise achiral system appears to fit the criteria for a chirodiastaltic interaction. Resolution can only be achieved in the presence of a large excess of one optical isomer

Table 1. Purity of the isolated enantiomers of [^{14}C-NCH_3]-nicotine.

	Total ^{14}C activity used (μCi)	Total ^{14}C activity isolated as pure enantiomer (μCi)	Radio-chemical purity (%)	Radio optical purity (%)
R-(+)-[^{14}C-NCH_3]-nicotine	75μCi	20μCi	96.7	99.3
S-(−)-[^{14}C-NCH_3]-nicotine	72μCi	24μCi	97.1	98.9

of nicotine. It seems likely that more than one physico-chemical property (solubility in the mobile phase, absorption onto the column packing, etc.) may be involved. The net result is that the lesser enantiomer fits less favorably into a mobile phase in which the order is built around its antipode; thus it has a longer retention time on the HPLC column and is effectively resolved.

Although the above hypothesis is speculative, it would appear to be the best available explanation. The possibility of association of enantiomers on a smaller scale (e.g., dimers, oligomers, etc.) can also be considered. This form of differential enantiomeric association between like and unlike pairs of enantiomers would require that 'dimers' formed from like and unlike pairs of enantiomers have different physical properties. This is possible, but no evidence exists of small scale association of nicotine molecules, either in solution or as a pure liquid.

This study represents the first reported preparation of enantiomerically pure ^{14}C-labeled nicotines. Using this system we have isolated large enough quantities of the chemically and optically pure enantiomers of radiolabeled nicotines for use in metabolic studies.

BIOMETHYLATION OF NICOTINE IN THE GUINEA PIG

The determination of nicotine and its metabolites is of particular interest in studying the correlation between the biological effects of tobacco smoking and the pharmacodynamics and pharmacokinetics of nicotine, the principal pharmacologically active component of tobacco.

Previous studies have shown that nicotine is primarily metabolized in humans to cotinine (3) and nicotine-N′-oxide (4) (Bowman et al., 1959; Gorrod and Jenner, 1975), both metabolites being formed via independent oxidation pathways. However, the stereospecific biomethylation of nicotine has also been reported (Cundy and Crooks, 1984; Cundy et al., 1985a,b), and recent studies indicate that N-methylnicotinium ion (5), the major methylated urinary metabolite of nicotine, is formed from a S-adenosylmethionine-dependent methyltransferase enzyme system that is widely distributed in tissues (Cundy et al., 1985 b,c; Cundy and Crooks, 1985; Damani et al., 1986). N-Methylnicotinium ion has the potential to function as an *in vivo* methylating agent (Pool and Crooks, 1985), and there is evidence that this metabolite is subsequently biotransformed *in vivo* in both mouse and rat, to nicotine (Aceto et al., 1983). This metabolic process must involve an N-demethylation step which, on mechanistic considerations, most likely involves direct displacement of an N-methyl group by some endogenous nucleophile. Work in our laboratory has therefore focused on examining in detail the metabolic pathways that operate in the biotransformation of nicotine to quaternary ammonium N-methylated metabolites.

The following studies on the *in vitro* and *in vivo* metabolism of radiolabeled nicotine enantiomers in the male Hartley guinea pig have utilized exclusively flow through radiochemical methodology for the identification and quantitation of nicotine metabolites in biological fluids, and have provided important new information on the comparative biotransformation of nicotine enantiomers to N-methylated products.

IN VITRO METABOLISM OF RADIOLABELED NICOTINES

We initially observed a chirodiastaltic effect in the analysis of *in vitro* metabolites of ^{14}C-(±)-nicotine in the guinea pig, which strongly suggested that nicotine was stereospecifically biotransformed into its N-methylated metabolite (see Fig. 5). RS-[^{14}C-2′]-Nicotine was incubated with dialyzed guinea pig spleen and other dialyzed tissue homogenates and the resulting biotransformation products identified and quantitated using radiochromatography as described previously. Figure 5a clearly shows that [^{14}C-2′]-nicotine gives rise to a single *in vitro* cationic metabolite with a retention on

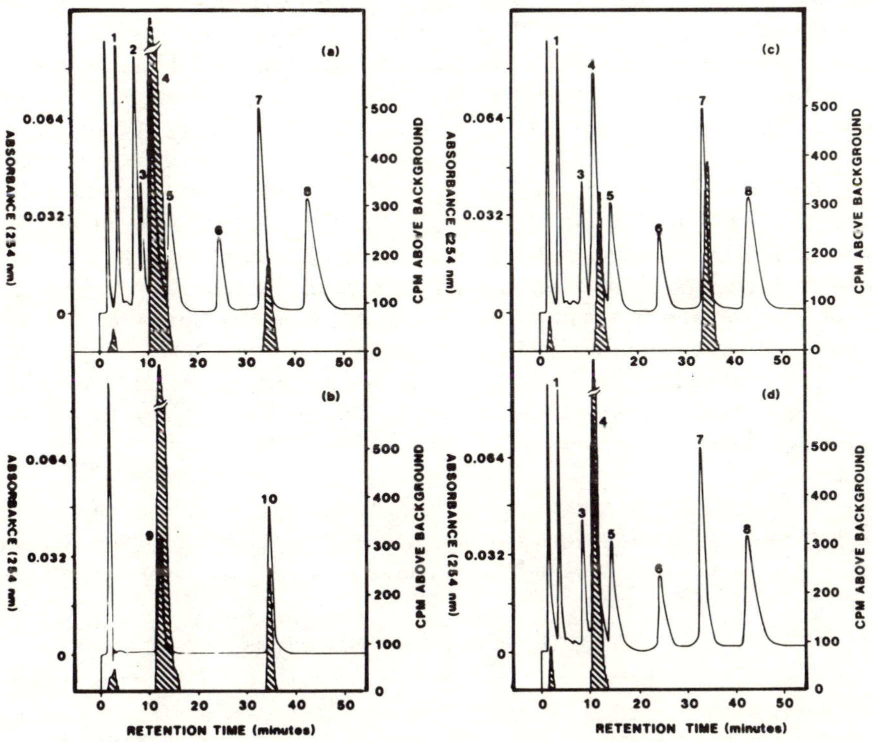

Figure 5. Radiochromatograms from analysis of *in vitro* homogenates by HPLC. Methylation of [^{14}C-2′]-*RS*-nicotine by dialysed spleen a) coinjected with *S*-N-methylnicotinium standard, b) coinjected with *R*-N-methylnicotinium standard. Stereospecific methylation of nicotine by dialysed lung using c) [^{3}H-NCH$_3$]-*R*-nicotine and d) [^{3}H-NCH$_3$]-*S*-nicotine. The standards used are (1) *S*-cotinine, (2) *S*-nornicotine, (3) *S*-nicotine-1′-oxide, (4) *S*-nicotine, (5) *S*-N-methylcotininium iodide, (6) *S*-N′-methylnicotinium iodide, (7) *S*-N-methylnicotinium iodide, (8) *S*-N, N′-dimethylnicotinium diiodide, (9) *R*-nicotine, (10) *R*-N-methylnicotinium iodide. The HPLC system is described elsewhere (Cundy et al., 1984a).

the HPLC system similar, but *not identical to,* that of the coinjected S-N-methylnicotinium ion standard; the radioactive metabolite eluted slightly later than the standard. At this stage, it was not possible to identify this metabolite as N-methylnicotinium ion. However, when the authentic standard of S-N-methylnicotinium iodide (5a) was replaced with R-N-methylnicotinium iodide (5b), and the same metabolic sample reanalyzed under identical chromatographic conditions, chromatogram b (Fig. 5) was obtained showing the radiolabeled cationic metabolite coeluting with an identical retention time to the R-N-methylnicotinium iodide standard.

It therefore appeared very likely that the same chirodiastaltic phenomenon observed with radiolabeled nicotine enantiometers was being exhibited by the radiolabeled N-methylnicotinium ion formed from the *in vitro* biotransformation of RS-[^{14}C-2′]-nicotine, and suggested that the metabolite was the stereochemically pure R-isomer of the N-methylnicotinium salt. This was subsequently confirmed, when the *in vitro* experiments were repeated with chirally pure S-(-)-[N′-C^{3}H$_3$]- and R-(+)- [N′-C^{3}H$_3$]-nicotine, respectively, in place of the racemic radiolabeled nicotine (see Figs. 5c and 5d respectively). As can be clearly seen, only the R-(+)-enantiomer was N-methylated by dialyzed guinea pig spleen homogenates, thus confirming the results from the chirodiastaltic effects obtained in the *in vitro* experiments with (±)-[^{14}C-2′]-nicotine. Other tissue homogenates also exhibited this stereospecificity in the N-methylation of the R-(+)-enantiomer of nicotine (see Table 2), indicating the presence and wide distribution of a 'nicotine N-methyltransferase' enzyme of high activity in the guinea pig. Dialyzed tissue homogenates invariably afforded greater N-methyltransferase activity when compared to undialyzed homogenates. Recent data (Cundy and Crooks, unpublished data) strongly indicate the presence of a heat-stable, small molecular weight endogenous compound capable of inhibiting nicotine N-methylation in all the tissues examined (see Table 2).

Interestingly, the levels of N-methylation of (±)-nicotine obtained after incubation with dialyzed lung homogenates were significantly less than 50% of that observed for the pure R-(+)-enantiomer (Table 3), suggesting that while S-(−)-nicotine was not a substrate for the N-methyltransferase

Table 2. Methylation of [^{14}C-2′]-*RS*-nicotine by guinea pig tissue homogenates.

	% Methylation/mg protein*		
Tissue	Undialysed	Dialysed	Significant difference
Liver	0.82±0.37	2.60±1.12	Yes
Lung	1.82±0.47	4.08±0.55	Yes
Spleen	3.62±0.69	8.42±2.14	Yes
Brain	2.84±0.25	5.21±0.55	Yes

*Expressed as a percentage of total ^{14}C activity in injection sample per mg protein.
Values are the mean of 4 determinations ± S.E.
Evaluated by a paired student's t test (P>0.05).

Table 3. Stereospecificity of N-methylation of nicotine by guinea pig tissue*

Substrate	% of Methylation/mg protein
[^{3}H-NCH_3]-*R*-nicotine	46.2
[^{3}H-NCH_3]-*S*-nicotine	None detected
[^{14}C-2′]-*RS*-nicotine	11.7
[^{14}C-NCH_3]-*RS*-nicotine (+)-bitartrate	12.0

*Using dialysed lung homogenate and incubating for 1 hour at 37°C.
Values are expressed as a percentage of the total radioactivity in the injection sample per mg protein.

enzyme, it might be capable of perturbing the N-methylation of its optical antipode. This was found to be the case (see Table 4), thus uncovering a remarkable substrate-inhibitor effect of nicotine enantiomers towards the guinea pig lung N-methyltransferase. R-(+)-Nicotine exhibited a Km of 1.42×10^{-5}M for the above enzyme, whereas S-(−)-nicotine was a competitive inhibitor (Ki=6.25×10^{-5}M) of the N-methylation of R-(+)-nicotine. *In vitro* experiments carried out with S-adenosyl-[$^{14}CH_3$]-methionine established that SAM was the methyl donor in this biomethylation reaction, and as with almost all SAM-dependent methyltransferase reactions, S-adenosyl-L-homocysteine, a product of the enzymic reaction, was a potent competitive feedback inhibitor (see Table 4) of R-(+)-nicotine methylation.

Table 4 also shows that nicotinamide is not a substrate for the 'nicotine N-methyltransferase' whereas histamine is a good substrate for this enzyme. Both of these compounds are substrates for SAM-dependent N-methyltransferases that were considered as possible candidates for the 'nicotine N-methyltransferase' enzyme. Histamine N-methyltransferase is an SAM-dependent enzyme present in the cytoplasm of most organs, which catalyses the N-methylation of histamine to N^{τ}-methylhistamine, as depicted in Scheme 1. The enzyme is widespread in mammals and is responsible for one of the major biotransformations of histamine (Schayer, 1974). From the data in Table 5, it is apparent that many of the properties of this enzyme bear a striking resemblance to those of 'nicotine N-methyltransferase'. Despite the difference in ring structure between histamine and nicotine, both substrates are predominantly in a monocationic form when methylated, unlike

Table 4. Substrate and inhibitor properties of guinea pig lung aromatic azaheterocycle N-methyltransferase

Compound	Substrate (K_m)	Inhibitor (K_i)
R-(+)-nicotine	14.2μM	—
S-(−)-nicotine	—	62.5μM
S-Adenosyl-L-methionine	13.2μM	—
S-Adenosyl-L-homocysteine	—	32.2μM
Nicotinamide	—	417.8μM
S-(-)-cotinine	—	N.D.[a]
Histamine	37.4μM	200μM

[a]not determined.

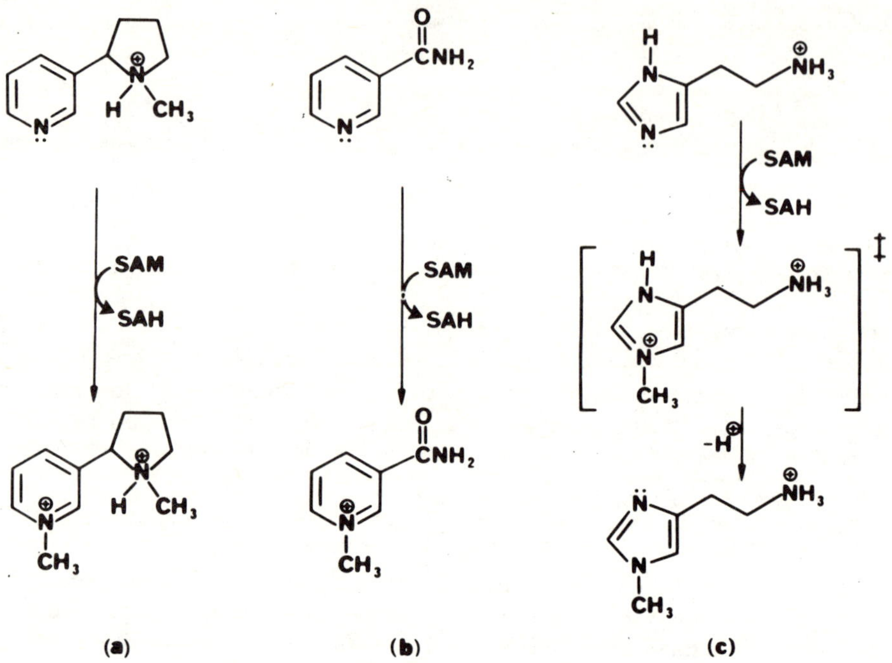

Scheme 1. Comparison of enzymic N-methylations: (a) N-methylation of R-(+)-nicotine to R-(+)-N-methylnicotinium ion; (b) N-methylation of nicotinamide to N′-methylnicotinamide by nicotinamide N-methyltransferase; (c) N-methylation of histamine to N^{τ}-methylhistamine by histamine N-methyltransferase.

Table 5. Comparison of the properties of the enzyme catalyzing R-(+)-nicotine N-methylation with those of histamine N-methyltransferase.

Property	R-(+)-Nicotine methyltransferase[a]	Histamine N-methyltransferase[a]
Substrate	R-(+)-Nicotine	Histamine
Methyl donor	SAM	SAM
Product	R-N-methylnicotinium ion	N^{τ}-methylhistamine
pKa of N-Atom methylated	3.04, 37°C	5.80, 37°C
Optimum pH	7.4	8.8
Ionic form of substrate at optimum pH	73.5% Monocation	79.9% Monocation
Organ distribution	Most organs	Most organs
Subcellular distribution	Cytoplasm	Cytoplasm
Effect of dialysis on activity	Increase	Increase
K_m of substrate[b]	1.42×10^{-5}M	4.3×10^{-5}M
K_m of SAM	1.32×10^{-5}M	1.6×10^{-5}M
K_i of SAH	6.25×10^{-5}M	0.4×10^{-5}M
Effect of ^{3}H-Nicotinamide	Not a substrate	Not a substrate
Effect of ^{3}H-Histamine	Substrate	Substrate

[a]Cundy, 1985 and references cited therein

[b]Refers to the substrate to which the methyl group is transferred.

nicotinamide, which is unionized. The ultimate product of histamine methylation is not a quaternary ammonium compound, but it is possible to envisage a mechanism for this biotransformation, which involves a transient quaternary ammonium intermediate (Scheme 1). The mechanism of this enzymic reaction has recently been investigated (Asano et al., 1984) by demonstrating a classsical S_N2 attack of the basic nitrogen atom of histamine on the methyl group of SAM, using a chiral methyl label. This study confirms that the initial methyl transfer step leads to a transient quaternary ammonium intermediate, which subsequently loses a proton. Thus, the N-methylation of nicotine and histamine may both involve very similar structures at the critical transition state of the enzymic reaction.

S-(−)-Nicotine is known to inhibit the N-methylation of histamine *in vitro* (Taylor and Lieber, 1979). This observation together with the above data seem to suggest that the same enzyme, i.e., histamine N-methyltransferase, may be implicated in the N-methylation of both nicotine and histamine. We are currently investigating the role of histamine N-methyl transferase in the biotransformation of nicotine to N-methylated products.

IN VIVO METABOLISM OF RADIOLABELED NICOTINE IN THE GUINEA PIG

We have also observed chirodiastaltic effects in the *in vivo* N-methylation of (±)-[^{14}C-NCH_3]-nicotine in the guinea pig (see Fig. 6). Here again, a radiochromatogram of the resulting urinary metabolites after I.P. administration of (±)-[^{14}C-NCH_3]-nicotine, showed a well retained metabolite band which did not coelute exactly with a UV-absorbing standard of S-(−)-N-methylnicotinium iodide. (Fig. 6a). However, when the above standard was replaced with R-(+)-N-methylnicotinium iodide, radiolabeled metabolite and standard coeluted with the same retention time (Fig. 6b). Figures 6c and 6d illustrate the results obtained with (±)-[^{14}C-2′]-nicotine; the extra metabolite eluting at 32 minutes is N-methylnornicotinium salt (6), a minor N-methylated nicotine metabolite (Sato and Crooks, 1985). Its absence in the radiochromatograms illustrated in Figs. 6a and 6b is due to loss of radiolabel as a consequence of N′-demethylation.

Stereospecific *in vivo* N-methylation of nicotine was verified by carrying out metabolic experiments with pure radiolabeled enantiomers. These studies proved conclusively that only the R-(+)-nicotine enantiomer was N-methylated *in vivo* in the guidea pig. In addition, a number of other N-methylated metabolites of R-(+)-nicotine could be detected. These metabolites have recently been isolated, identified, and quantified. The results are summarized in Scheme 2 and Table 6, which illustrate that, in the case of R-(+)-nicotine, both oxidation and N-methylation pathways of metabolism exist, whereas with the S-(−)-isomer only the oxidation pathway operates.

The biomethylation of R-(+)-nicotine appears to be a major route of metabolism for this enantiomer, yielding as much as 17 percent of recovered

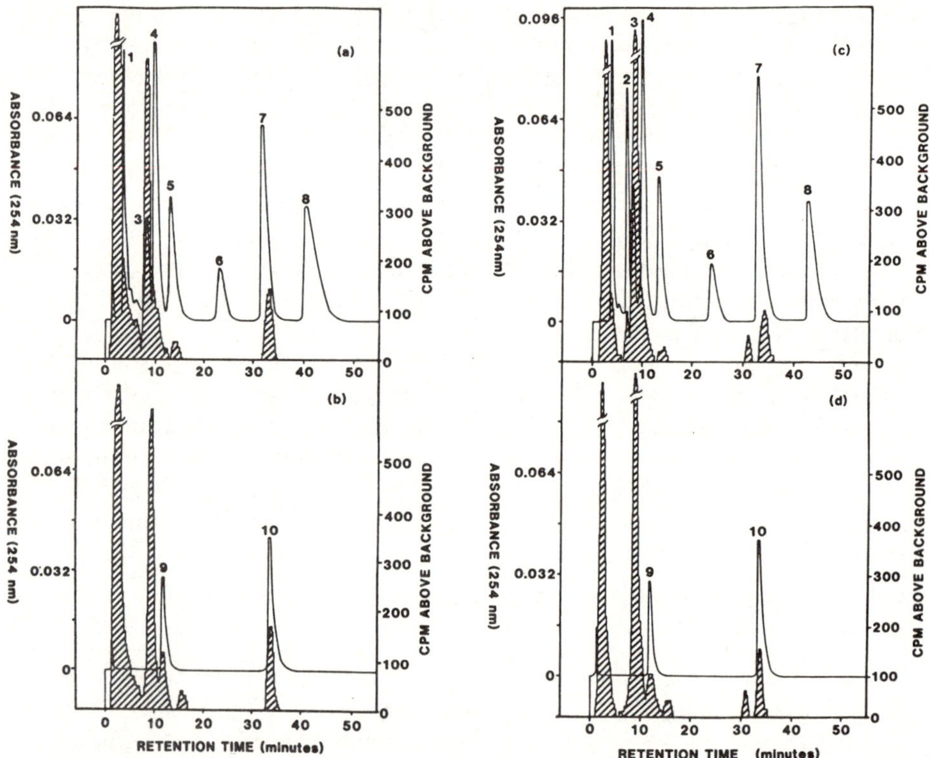

Figure 6. HPLC analysis of urine from guinea pigs administered ^{14}C-labeled nicotines. Methylation of [^{14}C-NCH_3]-(±)-nicotine-(+)-bitartrate in 10 hour urine (40μl); (a) coinjected with S-N-methylnicotinium standard (b) coinjected with *R*-N-methylnicotinium standard. Methylation of [^{14}C-2′]-(±)-nicotine in 2.5 hour urine (40μl); (c) coinjected with *S*-N-methylnicotinium standard and (d) coinjected with R-N-methylnicotinium standard. The standards used are (1) *S*-cotinine, (2) *S*-nornicotine, (3) S-nicotine-1′-N-oxide, (4) *S*-nicotine, (5) S-N-methylcotininium iodide, (6) S-N′-methylnicotinium iodide, (7) *S*-N-methylnicotinium iodide, (8) *S*-N, N′-dimethylnicotinium diiodide, (9) *R*-nicotine, (10) *R*-N-methylnicotinium iodide. Column: Partisil-10 SCX (25 cm x 4.6 mm I.D.); Eluent: (A) 0.3M sodium acetate/methanol (70:30), pH 4.5, (B) 0.3M sodium acetate/methanol (70:30) + 1% triethylamine, final pH 4.5; Gradient; 100% A until 12 minutes, rising to 100% B over 10 minutes; Flow rate: 2.0 ml/min.

methylated urinary metabolites. These metabolites have been identified as N-methylated products of nicotine, cotinine and nornicotine, together with a minor and as yet unidentified N-methylated metabolite (Pool and Crooks, 1985). It is interesting to note that both *in vivo* and *in vitro* biotransformations of R-(+)-nicotine afford only metabolites that carry a methyl group on the pyridinium nitrogen; no pyrrolidinium N-methyl metabolites of R-(+)-nicotine have been detected in these studies. This regiospecificity is puzzling, when one considers the relative basicity of the two nitrogen atoms in the nicotine molecule. The pyridine nitrogen (pKa = 3.04) is much less basic than the pyrrolidine nitrogen (pKa = 7.84). Thus N-methylation oc-

Table 6. Quantification of methylated metabolites in 24 hour urine after I.P. administration of radiolabeled R-(+)- and S-(−)-nicotines, and R-(+)-N-methylnicotinium acetate in male Dunkin Hartley guinea pigs.

Radiolabeled compound	Total recovery of radiolabel	% Recovery of radiolabeled metabolite in 24 h urine[a]						
		Cotinine and polar metabolites	NNO[b]	Nicotine	NMC[c]	NMNor[d]	NMN[e]	New metabolite[f]
S-(−)-[^{3}H-NCH_3]-nicotine	73.46 ±3.81	76.0 ±2.02	19.96 ±0.90	3.40 ±1.31	0	0	0	0
R-(+)-[^{3}H-NCH_3]-nicotine	51.82 ±0.86	72.5 ±4.87	14.9 ±4.25	0	2.42 ±0.48	3.2[g] ±0.2	9.86 ±0.97	1.20 ±0.30
R-(+)-[^{14}C-NCH_3]-N-methylnicotinium acetate	61.9 ±14.3	4.3 ±1.1	—[h]	—[h]	17.5 ±0.8	8.0 ±0.7	60.9 ±1.5	9.0 ±0.5

[a] Values given are mean ±SE,n=5;
[b] nicotine N′-oxide;
[c] N-methylcotininium ion;
[d] N-methylnornicotinium ion;
[e] N-methylnicotinium ion;
[f] tentatively identified as N-methyl-N′-oxonicotinium ion;
[g] value based upon that obtained with (±)-[^{14}C-2′]-nicotine (Cundy and Crooks, 1984; Cundy et al. 1985b);
[h] not determinable due to loss of radiolabel.

curs at the least basic N-atom, even though the reaction is catalyzed by an SAM-dependent methyltransferase (Scheme 1), which presumably operates via a classical S_N2 nucleophilic displacement mechanism.

The presence of N-methylcotininium (7) and N-methylnornicotinium (6) ions in the urine of guinea pigs treated with both R-(+)-nicotine and R-(+)-N-methylnicotinium acetate (Pool and Crooks, 1985) clearly demonstrates that nicotine N-methylation can initially occur *in vivo,* followed by oxidation or N′-demethylation to further metabolites. Nevertheless, the alternate pathway of oxidation (or N′-demethylation) followed by N-methylation cannot be ruled out. In this latter respect, an early study by McKennis et al. (1963) has shown that cotinine can be N-methylated *in vivo* in man and dog.

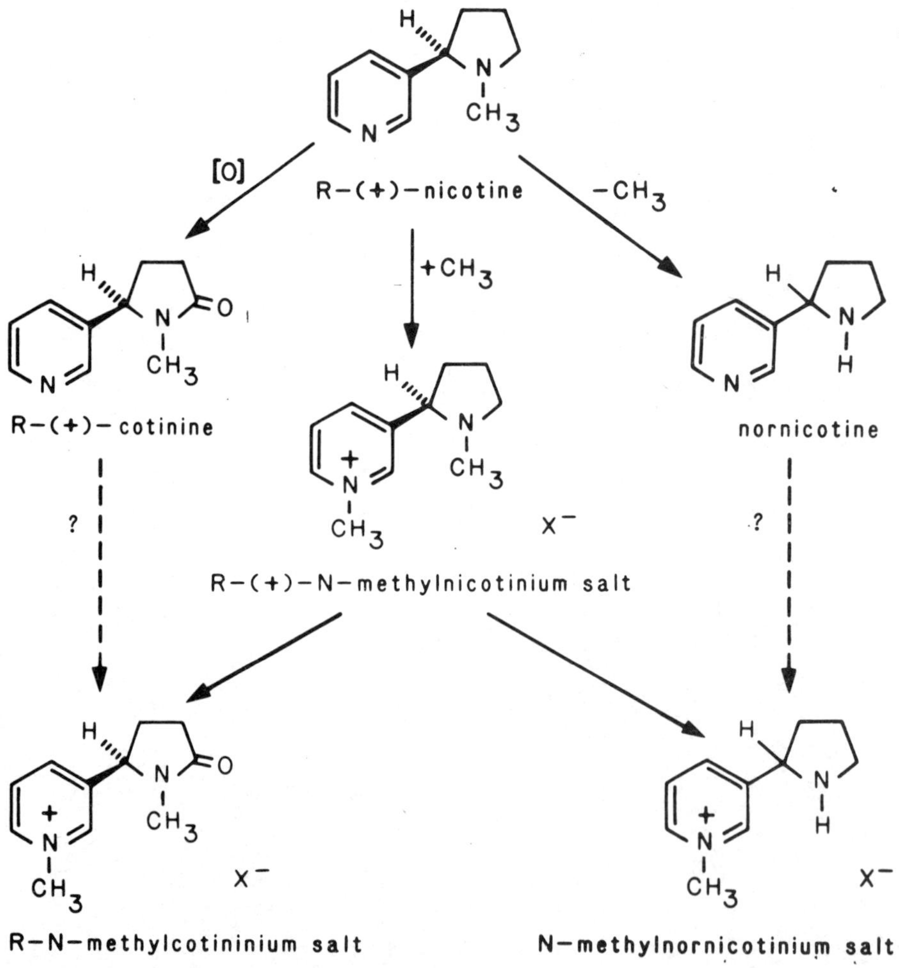

Scheme 2. Proposed pathways for the *in vivo* formation of N-methylated metabolites of R-(+)-nicotine in the guinea pig.

It is interesting to speculate on the identity of the unkown minor metabolite (see Table 6), which was detected in the urine of guinea pigs that had been treated with both R-(+)-[^3H-N'-CH_3] nicotine and R-(+)-[^{14}C-NCH_3]-methylnicotinium acetate. The metabolite exhibited long retention characteristics on cation-exchange chromatography and is almost certainly a quaternary amine. It must contain both N-methyl groups intact and is probably a secondary N-methylated nicotine metabolite. The radiolabeled metabolite did not coelute with either R-(+)-N'-methylnicotinium iodide or R-(+)-N,N'-dimethylnicotinium diiodide. A possible candidate for this metabolite is N-methyl-N'-oxonicotinium ion (8). Work is in progress to determine the identity of this new nicotine metabolite.

These data from our metabolic radiotracer studies have established that N-methylnicotinium ion plays a central role in the *in vivo* methylation pathway for R-(+)-nicotine in the guinea pig (see Scheme 2). The ability of both R-(+)- and S-(−)-isomers of nicotine to interact, either as a substrate or as an inhibitor, with an as yet unidentified S-adenosylmethionine-dependent methyltransferase system raises the question of whether exposure to 'smoke nicotine', which is known to contain both enantiomers of nicotine, may modify endogenous methyltransferase activity or result in alterations of tissue levels of SAM, and/or SAH. Significant changes in the levels of the latter two endogenous compounds could potentially lead to an alteration of tissue methylation activity, which in turn could result in a perturbation of the levels of important biogenic amines that are normally metabolized by SAM-dependent methyltransferase enzymes. In addition, the potential of N-methylated products of nicotine metabolism to act as methylating agents *in vivo* requires further study. We are currently investigating the role of nicotine methylation in the toxicology of tobacco products.

ACKNOWLEDGMENT.

This study was supported by the Tobacco and Health and Research Institue, Lexington, Kentucky (research grants 4C018 and 5C31 to PAC).

REFERENCES

Aceto, M.D., Awaya, H., Martin. B.R. and May, E.L. (1983). Antinociceptive action of nicotine and its methiodide derivatives in mice and rats. *Brit. J. Pharmacol.* 79, 869-876.

Anderson, A., Hansson, E. and Schmitterlow, C.A. (1965). Gastric excretion of ^{14}C-nicotine. *Experientia* 21, 211-213.

Appelgren, L.E., Hansson, E. and Schmitterlow, C.A. (1962). The accummulation and metabolism of ^{14}C-labeled nicotine in the brain of mice and rats. *Acta Physiol. Scand.* 56, 249-257.

Asano, Y., Woodard, R.W., Houck, D.R. and Floss, H.A. (1984). Stereochemical course of the transmethylation catalysed by histamine N-methyltransferase. *Arch. Biochem. Biophys.* 231, 253-256.

Bowman, E.R., Turnbull, L.B. and McKennis, Jr.,H. (1959). Metabolism of nicotine in the human and excretion of pyridine compounds by smokers. *J. Pharmacol. Exp. Ther.* 127, 92-95.

Cundy, K.C. and Crooks, P.A. (1983). Unexpected phenomenon in the high-performance liquid chromatographic analysis of racemic ^{14}C-labeled nicotine: separation of enantiomers in a totally achiral system. *J. Chromatogr.* 281, 17-33.

Cundy, K.C. (1984). N-Methylation of nicotine enantiomers in the guinea pig. *Ph. D. Dissertation, College of Pharmacy, University of Kentucky.*

Cundy, K.C. and Crooks, P.A. (1984). High-performance liquid chromatographic method for the determination of N-methylated metabolites of nicotine. *J. Chromatogr. Biomed. Appl.* 306, 291-301.

Cundy, K.C., Godin, C.S. and Crooks, P.A. (1984). Evidence of stereospecificity in the *in vivo* methylation of [^{14}C]-(±)-nicotine in the guinea pig. *Drug Metab. Disposit.* 12, 755-759.

Cundy, K.C. and Crooks, P.A. (1985). *In vitro* characteristics of guinea pig lung aromatic azaheterocycle N-methyltransferase. *Drug Metab. Disposit.* 13, 658-663.

Cundy, K.C., Crooks, P.A. and Godin, C.S. (1985a). Remarkable substrate-inhibitor properties of nicotine enantiomers towards a guinea pig lung aromatic azaheterocycle N-methyltransferase. *Biochem. Biophys. Acta* 128, 312-316.

Cundy, K.C., Godin, C.S. and Crooks, P.A. (1985b). Stereospecific *in vitro* N-methylation of nicotine in guinea pig tissue by an S-adenosylmethionine-dependent N-methyltransferase. *Biochem. Pharmacol.* 34, 281-284.

Cundy, K.C., Sato, M. and Crooks, P.A. (1985c). Stereospecific *in vivo* N-methylation of nicotine in the guinea pig. *Drug Metab. Disposit.* 13, 175-185.

Damani, L.A., Shaker, M.S., Godin, C.S., Crooks, P.A., Ansher, S.S. and Jakoby, W.B. (1986). The ability of arylamine N-methyltransferase to N-methylate azaheterocycles. *J. Pharm. Pharmacol.* 38, 547-550.

Decker, K. and Sammeck, R. (1964). On the degradation of nicotine in cell-free liver preparations. *Deutsch. Med. Forsch.* 2, 33-36.

Gorrod, J.W. and Jenner, P. (1975). Metabolism of tobacco alkaloids. *Essays in Toxicol.* 6, 35-78.

Hansson, E. and Schmitterlow, C.A. (1962). Physiological disposition and fate of ^{14}C-labeled nicotine in mice and rats. *J. Pharmacol. Exp. Ther.* 137, 91-102.

Horeau, A. and Guette, J.P. (1974). Diastereomeric interactions of antipodes in the liquid phase. *Tetrahedron* 30, 1923-1931.

Jacob, III, P. (1982). Resolution of (±)-5-bromonornicotine. Synthesis of (R)-and (S)-nornicotine of high enantiomeric purity. *J. Org. Chem.* 47, 4165-4167.

Kabachnik, M.I., Mastryukova, T.A., Fedin, E.I., Vaisberg, M.S., Morozov, L.L., Petrosky, P.V. and Shipov, A.E. (1976). An NMR study of optical isomers in solution. *Tetrahedron* 32, 1719-1728.

Klus, H. and Kuhn, H. (1977). A study of the optical activity of smoke nicotines. *Fachliche Mitt. Oesterr. Tabakregie* 17, 331-334.

McKennis, Jr., H., Wada, E., Bowman, E.R. and Turnbull, L.B. (1961). Demethylation in the metabolism of (−)-nicotine *in vivo*. *Nature* (London) 190, 910-911.

McKennis, Jr., H., Turnbull, L.B. and Bowman, E.R. (1963). N-methylation of nicotine and cotinine *in vivo*. *J. Biol. Chem.* 238, 719-723.

Pool, W.F. and Crooks, P.A. (1985). Biotransformation of primary nicotine metabolites. 1. *In vivo* metabolism of R-(+)-[^{14}C-NCH_3]-N-methylnicotinium ion in the guinea pig. *Drug Metab. Disposit.* 13, 578-581.

Sato, M. and Crooks, P.A. (1985). N-Methylnornicotinium ion, a new *in vivo* metabolite of R-(+)-nicotine. *Drug Metab. Disposit.* 13, 348-353.

Schayer, R.W. (1974). Histamine methylation *in vivo* and *in vitro*. *Agents and Actions* 4/3, 185-186.

Schmitterlow, C.A. and Hansson, E. (1962). Physiological disposition and fate of nicotine labelled with carbon-14 in mice. *Nature* (London) 194, 298-299.

Taylor, S.L. and Lieber, E.R. (1979). *In vitro* inhibition of rat intestinal histamine-metabolizing enzymes. *Food Cosmet. Toxicol.* 17, 237- 240.

Williams, T., Pitcher, R.A., Bommer, P., Gutzwiller, J. and Uskokovic, M. (1969). Detection of chirality with the chemical ionization mass spectrometer. *J. Amer. Chem. Soc.* 91, 1871-1872.

Progress in HPLC, Vol. 3, pp. 149-165
Parvez et al. (Eds)

A simplified method for determination of radioactive metabolites of benzo(a)pyrene using HPLC with radioactive flow detection

TERESA GESSNER and JANUSZ Z. BYCZKOWSKI

Department of Experimental Therapeutics, Roswell Park Memorial Institute, New York State Department of Health, Buffalo, NY 14263, USA

SUMMARY

The use of radioflow detector for quantitation of ^{14}C or ^{3}H labeled BaP metabolites was investigated. Isolated mouse liver microsomes were employed for generation of oxidized and glucuronidated metabolites of B(a)P. Free radical oxidation products of B(a)P were also examined. A method for rapid extraction, separation and quantitation of oxidized metabolites of B(a)P and their glucuronate conjugates is described. Use of radioflow detector in series with HPLC system reduced the analysis time approximately to the duration of the chromatographic run. This isotopic method approximated the sensitivity of the conventional liquid scintillation counting of fractions; the observed counting efficiencies were in the range of 30% for ^{3}H and 70% for ^{14}C.

INTRODUCTION

Benzo(a)pyrene is a widespread environmental pollutant, known to cause cancer in experimental animals and most probably in man. Its fate, disposition and metabolism have been extensively investigated at the level of the whole animal (Falk *et al.*, 1962; Daniel *et al.*, 1967; Raha, 1972), isolated and perfused organs such as liver (Smith and Bend, 1979; Sweeny and Reinke, 1986), or lung (Smith *et al.*, 1978; Ball *et al.*, 1979; Smith *et al.*, 1980; Bend *et al.*, 1981; Warshawsky *et al.*, 1981; Warshawsky *et al.*, 1983; Warshawsky *et al.*, 1984), various tissue homogenates (Sims, 1967; Bond, 1983; Nishimoto and Varnasi, 1985), isolated cells *in vitro* (Jones, *et al.*, 1978; Zaleski *et al.*, 1983), or in culture (Gelboin *et al.*, 1976; Cohen *et al.*, 1976; Autrup *et al.*, 1978; Vaught *et al.*, 1978; Marshall *et al.*, 1979; Selkirk *et al.*, 1980; Moore and Gould, 1984; Sadowski *et al.*, 1985; Merrick and Selkirk, 1985), as well as at the level of isolated microsomes (Selkirk *et al.*, 1974; Holder *et al.*, 1975; Yang, *et al.*, 1977; Selkirk, 1980;

Murphy and Hecht, 1985; Haake *et al.*, 1985), and other organelles (Bansal *et al.*, 1981; Oesch *et al.*, 1985), isolated, purified and/or reconstituted enzymes (Thakker *et al.*, 1978; Reed and Marnett, 1982), and finally at nonenzymatic catalytically-active biomolecules (Dix *et al.*, 1985), and in the environment (Pitts *et al.*, 1980).

The metabolites of benzo(a)pyrene directly found or deduced from the mechanism of the metabolic pathway in those systems include:

1.) primary epoxides, mostly converted to monohydroxyphenols and dihydro-diols;
2.) diolepoxides, converted to triols and tetrols;
3.) diones and resultant quinols;
4.) conjugates of these oxygenated metabolites with glucuronic acid, sulfate and glutathione.

An excellent review of the mechanisms of metabolite formation, the enzymes involved therein, and the stereochemical aspects of their formation was written by Gelboin (1980).

Production of oxygenated intermediates, leading to the production of 7,8-diol-9,10-epoxide-2, is considered to be the means of metabolic activation of benzo(a)pyrene, as that compound binds covalently to cellular macromolecules and was pronounced "an ultimate carcinogen" (for review see Thakker *et al.*, 1985). By contrast, conjugation of the oxygenated metabolites with glucuronic acid, sulfate or glutathione is usually regarded as a detoxication process, which leads to water-soluble products that can be excreted (Ashurst *et al.*, 1979; see also Zaleski *et al.*, 1983 and references therein). The overall carcinogenic effect of benzo(a)pyrene depends, therefore, on a balance between activation and deactivation reactions.

From the analytical point of view, the investigation of this balance in a mixture of benzo(a)pyrene metabolites produced by any biosystem represents a real challenge, since the separation of derivatives, with often very similar physical and chemical properties, is difficult and usually incomplete (Weigert and Mottram, 1946). Apart from classical extraction procedure, several other methods and techniques were tried to resolve analytically the metabolic mixtures of benzo(a)pyrene derivatives:

1.) Chromatography on alumina column (Cook *et al.*, 1950), which, using proper solvents gave separation of nonconjugated metabolites from sulfate, glucuronide and glutathione-conjugates (Autrup, 1979);
2.) Thin-layer chromatography, which, combined with fluorescence detection permitted the separation and quantitation of nonconjugated products of benzo(a)pyrene oxygenation (Waterfall and Sims, 1972);
3.) Sephadex column chromatography, which, combined with liquid scintillation counting of collected fractions, permitted the separation and quantitation of DNA adducts with radioactively/labeled benzo-(a)pyrene metabolites (Sims *et al.*, 1974);
4.) Reversed-phase thin-layer chromatography, which, with the addition of an ion-pairing reagent afforded the separation of major ben-

zo(a)pyrene metabolites, including sulfate and glucuronide conjugates (Marshall *et al.*, 1980); and

5.) Gas chromatography, which, in combination with a mass-spectrometer, allowed the separation and detection of some benzo(a)pyrene phenols, triols and tetrols (Takahashi *et al.*, 1979).

However, a real breakthrough in analysis of benzo(a)pyrene metabolites was achieved in 1974 by application of high pressure liquid chromatography, known as high performance liquid chromatography - HPLC (Selkirk *et al.*, 1974a; for review see Thakker *et al.*, 1978 and references therein). Using reversed-phase HPLC with fluorescence or UV spectrophotometric detection, in combination with liquid scintillation measurement of radioactively labeled benzo(a)pyrene metabolites in collected fractions, the separation and identification of eight major benzo(a)pyrene unconjugated metabolites (Holder *et al.*, 1974; Selkirk et al., 1974), as well as characterization of the effect of microsomal enzyme-induction on metabolite pattern (Yang *et al.*, 1975; Holder *et al.*, 1975) were accomplished. The eight major benzo(a)pyrene metabolites included: 9,10-diol; 4,5-diol; 7,8-diol; 1,6-dione; 3,6-dione; 6,12-dione, 9-monohydroxy and 3-monohydroxy derivatives. The ninth, quite labile metabolite, separated by HPLC and characterized was 4,5-epoxy-benzo(a)pyrene (Selkirk *et al.*, 1975; Lam and Wattenberg, 1977). Improvements in HPLC technique allowed also to separate tetrols and triols (Yang *et al.*, 1977; Keller *et al.*, 1982; Dix and Marnett, 1984), separate and characterize adducts of diol epoxides with nucleosides and DNA (Thakker *et al.*, 1978a; Nishimoto and Varnasi, 1985), and even resolve the enantiomers of the diols (Thakker *et al.*, 1977; Weems and Yang, 1982). Due to the high sensitivity and reliability of the radioactivity measurement, it was possible to develop a dual-label HPLC assay, which separated and quantitated as low as fentomolar amounts of benzo(a)pyrene metabolites in biological material (Murphy and Hecht, 1985).

In the present paper, a simple HPLC assay is described, which allows us to quantitate, in the same sample, both the water-soluble glucuronide conjugates and the hydrophobic oxygenation products of benzo(a)pyrene, generated by mouse liver microsomes *in vitro*. The simplification consists of both a facile single step extraction procedure and a rapid quantitation with the aid of a radioactive flow detector.

MATERIALS AND METHODS

Materials

[7,10-^{14}C] benzo(a)pyrene, specific radioactivity 58.5 mCi/mmol and G-^{3}H benzo(a)pyrene, specific radioactivity 76Ci/mmol were purchased from Amersham. Nonradioactive benzo(a)pyrene, NADPH, UDPGA, KO_2 and superoxide dismutase were obtained from Sigma Chemical Co. Liquiscint 2 complete liquid scintillation cocktail was purchased from National Diagnostics, and FLO-SCINT II complete nongelling scintillator was obtained from

Radiomatic Instruments and Chem. Co. Inc. Methanol, HPLC-grade was purchased from J.T. Baker Chem. Co. All other reagents used were of analytical or HPLC grade and were purchased from Fisher Sci. Co.

Preparation of microsomes

Mouse liver microsomes were prepared as previously described (Byczkowski and Gessner, 1987a).

Peroxidized microsomes were prepared from native ones by diluting 1 ml of the fresh microsomal suspension with an equal volume of the buffered medium and incubating for 10 minutes at 37°C in the presence of 0.1 mM $Fe_2(SO_4)_3$ and 0.1 mM Fe˙EDTA (pH 7.4) prepared *ex tempore* or Fe^{3+}˙FeEDTA with addition of 142 μg of KO_2. This amount of KO_2, equivalent to a final concentation of 1 mM, gave about 7% of reactive $0_2^{\cdot-}$ anion-radicals (Cunningham and Lokesh, 1983), so the initial concentration of about 70 μM of the damaging $0_2^{\cdot-}$ was produced in the peroxidative system, before it was trapped by subsequent reactions (Lesko et al., 1980). The Fe^{3+}˙FeEDTA-preincubation and Fe^{3+}˙FeEDTA *plus* $0_2^{\cdot-}$ -preincubation were terminated by adding ice-cold Tris-HCl buffer (pH 7.2), containing 154 mM KCl, up to the volume of 25 ml, and centrifuging for 45 minutes at 105,000 x g. Pellets were resuspended in the same medium as the native microsomes and protein concentration in peroxidized microsomes was set at 10 mg/ml.

Incubations

The incubations were done in buffered medium, final volume 0.2 ml, containing: 130 mM NaCl, 5.2 mM KCl, 1.3 mM KH_2HPO_4, 10 mM Na_2HPO_4, 1.3 mM $MgSO_4$ (pH 7.4), and when indicated, supplemented with 10 mM UDPGA, 5 mM NADPH and microsomal protein at concentration of 2.5 mg/ml. All components were mixed in silicone-coated, borosilicate-glass, stoppered test-tubes, and reaction was started by adding 5 μl or 10 μl of acetone solution of [^{3}H]-benzo(a)pyrene (0.76 μCi/incubation) or [^{14}C]-benzo(a)pyrene (0.585 μCi/incubation) for a final concentration of 50 μM. Zero time controls, containing the same additions mixed with boiled microsomes were kept on ice. Incubations were performed in a Dubnoff shaking water bath at 37°C for 10 minutes. After that time the reactions were terminated by addition of 0.3 ml of chilled methanol (for HPLC analysis) or 0.2 ml of 0.15M KOH in 85% DMSO (for total benzo(a)pyrene metabolite assay).

Measurement of protein-bound metabolites

After termination of the reaction with 0.3 ml of methanol, samples were centrifuged for five minutes at 1,000 x g. Supernatants were used for HPLC analysis whereas the pellets were washed with methanol until all washable radioactivity was gone (usually 4-5 times with 1 ml portions of methanol). Washed pellets were dissolved overnight at 40°C with 0.5 ml of 1.0M

NaOH, neutralized with 1.0 M HCl and radioactivity was counted in a Packard Liquid Scintillation Spectrometer using Liquiscint 2 complete scintillation fluid. Counting efficiency was 70-80% with [^{14}C] benzo(a)-pyrene.

Measurement of total metabolism of benzo(a)pyrene

Overall metabolism of benzo(a)pyrene was estimated in the incubation samples using the method of DePierre *et al.* (1975) with modification of Van-Cantfort *et al.* (1977). After termination of reaction with 0.2 ml of 0.15 M KOH in 85% DMSO the unmetabolized benzo(a)pyrene was extracted two times with 2.5 ml of hexane using a maximum speed of Super-mixer (Lab-Line Instruments Inc., Melrose Park, IL). Extracts were centrifuged for 5 minutes at 1,000 x g, 0.2 ml of water phase acidified with equal volume of 1.0 M HCl and radioactivity of the sample was counted by liquid scintillation. Counting efficiency was 30-40% with [^{3}H]benzo(a)pyrene.

HPLC analysis and quantitation of radioactive peaks

Supernatants from methanol-terminated incubation mixtures (after spinning down the precipitated proteins and salts) were injected directly into the HPLC. Thus, aliquots of 200 μl of these methanolic extracts were injected into the Perkin-Elmer Series 400 Liquid Chromatograph, equipped with 200 μl loop 7125 Rheodyne syringe loading sample injector, connected to Rheodyne in-line filter and Waters μBondapak C18 guard-pak, and finally to Waters μBondapak C18, (0.39 cm I.D. x 30 cm long) analytical column. The separation program used a flow rate at 1 ml/min and a linear water/methanol gradient as follows:

At step 0, before sample injection, the column was equilibrated for at least five minutes with 46% methanol; after injection of the analytical sample, step 1 lasted 45 minutes during which time methanol concentration increased with a linear gradient from 46% to 95%; this was followed by step 2, changing linearly the concentration of methanol during 10 minutes from 65% to 99%. During the last five minutes the column was washed with a rapidly decreasing methanol concentration from 99% to 46%.

The column eluent was continuously monitored with a Waters 490 multiwavelength detector set at 276 nm and, connected to a Perkin-Elmer R-100A recorder for detection of reference peaks, and with Radiomatic Instruments FLO-ONE/BETA Model IC radioactivity flow detector. The radioactivity detector was connected to the data processing system consisting of a Micromate PMC computer, Qume Co. monitor and C. Itoh Inc. electronic printer. The radioactivity detector system was set to monitor continuously the radioactivity in the ^{14}C or ^{3}H channels and absorbance at 276 nm at UV channel with 30 seconds delay (resulting from the time necessary for the eluent to flow through the tubing connecting the UV detector with the radioactivity detector).

RESULTS

An example of authentic, radioactivity tracings from HPLC analysis of free-radical-initiated oxidation products of [^{14}C] and [^{3}H]-benzo(a)pyrene by $O_2^{\cdot -}$-peroxidized mouse liver microsomes is shown in Figure 1. The retention times of standards separated under identical condition and on the same column were: 3-4 minutes $^{3}H_2O$, 19-20 minutes 7,8,9-triol, 25-26 minutes 4,5,-diol, 27-28 minutes 7,8 diol, 30-31 minutes 1,6-dione, 32-33 minutes 3,6-dione, 35-36 minutes 6,12-dione, 40-41 minutes 3-monohydroxy, and 46-47 minutes benzo(a)pyrene. The background subtraction was set to 50 cpm in both tracings (Fig. 1A and B). For integration purposes and further calculations higher background subtractions were chosen, especially for [^{3}H]-labeled products, and the minimum integrated peak area was routinely set to 300 cpm. This example illustrates the relative sensitivity of detection when the radioactivity is in the form of ^{14}C as compared to ^{3}H label, in approximately comparable amounts. The conditions of the radioactive flow detection system were as described below for the more complex example with many more closely eluting peaks.

An example of an HPLC profile of [^{14}C] benzo(a)pyrene metabolic oxygenation products by native mouse liver microsomes, incubated in the presence of NADPH, is shown in Figure 2. The metabolic pattern was compared with a series of standards (upper panel) and nonenzymatic oxidation products of [^{3}H] benzo(a)pyrene, produced by $O_2^{\cdot -}$-peroxidized mouse liver microsomes, incubated without NADPH (shaded area). It can be seen from Figures 1 and 2 that the multiple metabolites of benzo(a)pyrene produce many peaks, which elute only about a minute apart. To achieve good resolution and quantitation of these peaks in the radioactive flow detector, we found that we had to use a 0.5 ml flow cell and 1:3 ratio of the HPLC eluent: scintillation cocktail, while the total flow through the cell was restricted to 4 ml/min.

Thus, ^{14}C and ^{3}H were counted in the 0.5 ml cell, by mixing 1 ml/min of eluent with 3 ml/min of FLO-SCINT II scintillation cocktail. Using quench-curves prepared according to FLO-ONE/BETA radioactive flow detector manual the counting efficiencies ranged from 67% to 73% for ^{14}C and from 28% to 35% for ^{3}H through the methanol/water gradient. The stream splitter was bypassed and the splitter ratio set to 0. The minimum integrated peak area was set at 300 cpm, and 120 cpm background was routinely subtracted in the radioactivity channels.

The first polar peak, appearing in the pattern of [^{3}H] benzo(a)pyrene products (shaded area in Fig. 2), cochromatographed with $^{3}H_2O$ and increased with increasing overall amount of [^{3}H] benzo(a)pyrene oxidation products.

Among the products of nonenzymatic oxidation of [^{3}H] benzo(a)pyrene by $O_2^{\cdot -}$-peroxidized microsomes, incubated without NADPH, diones (6,12-dione) were predominant, no monohydroxy derivatives were detected, and only small amounts of the product which cochromatographed with triols

(Fig. 2, shaded area) were detectable. In contrast to that, 3-monohydroxybenzo(a)pyrene (Fig. 2) was the dominant oxygenated metabolite of [^{3}H] benzo(a)pyrene or [^{14}C] benzo(a)pyrene, produced by native microsomes with NADPH as the source of reducing equivalents. Other major metabolites identified by HPLC included: 6,12-dione, 3,6-dione, 1,6-dione, 7,8-diol, 4,5-diol, and 9,10-diol, and some triols and tetrols as minor metabolites.

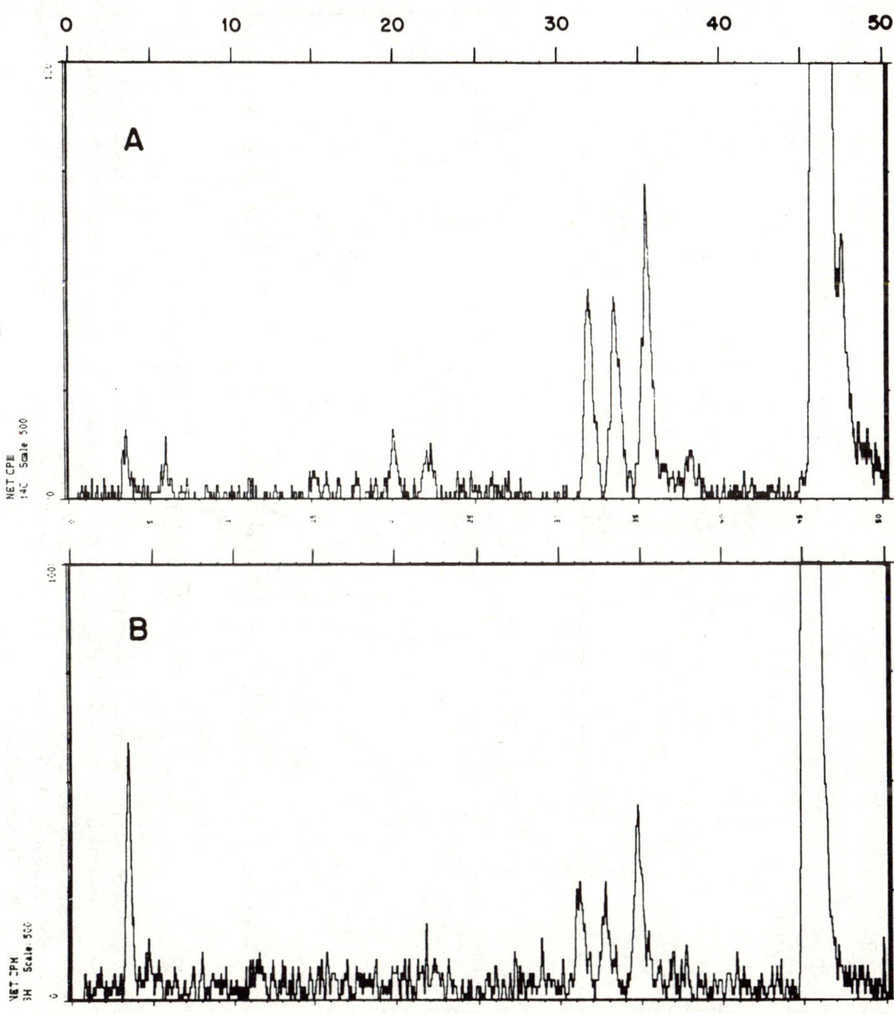

Figure 1. Authentic HPLC profiles of free-radical-initiated oxidation products of 50 μM benzo(a)pyrene by $O_2^{\cdot-}$-peroxidized mouse liver microsomes with NADPH. Panel A represents a pattern of [^{14}C] benzo(a)pyrene products (0.585 μCi/incubation) and panel B represents a pattern of [^{3}H] benzo(a)pyrene products (0.76 μCi/incubation). Radioactivity counts/min. (after subtraction of background set at 50 cpm) *vs.* retention time in minutes are shown. For experimental details see Materials and Methods.

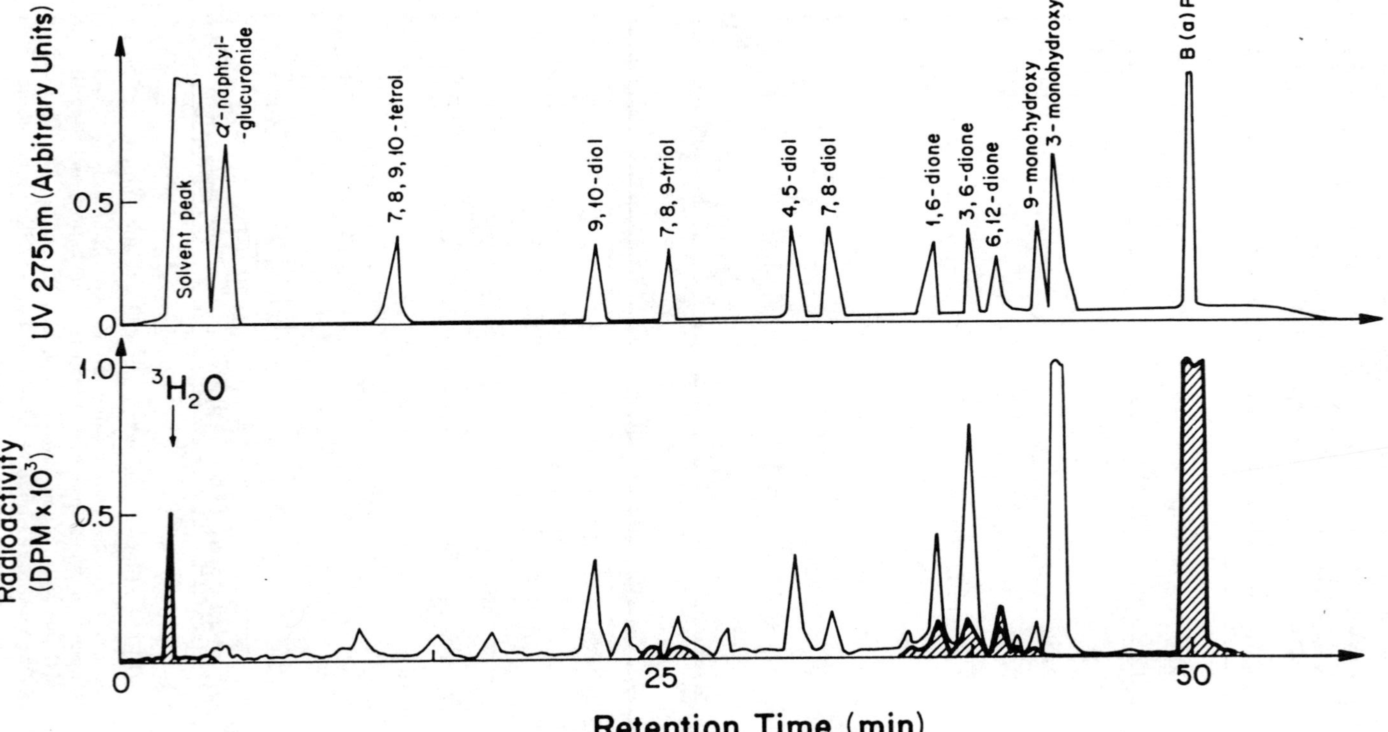

Figure 2. HPLC profiles of metabolic oxygenation and free-radical-initiated oxidation products of benzo(a)pyrene by mouse liver microsomes with NADPH and $0_2^{\cdot -}$-peroxidized microsomes without NADPH. Upper panel represents UV scan of standards. Native mouse liver microsomes were incubated, as described in Materials and Methods, in the presence of 5 mM NADPH and 50 μM [^{14}C] benzo(a)pyrene (0.58 μCi/incubation). Shaded area represents result of incubating $0_2^{\cdot -}$-peroxidized microsomes with 50 μM [^{3}H] benzo(a)pyrene (0.76 μCi/incubation) without NADPH. For experimental details see Materials and Methods.

In some experiments total metabolism of B(a)P was estimated in the incubation samples using the method of DePierre *et al.*, (1975) with modification of VanCantfort *et al.*, (1977). Briefly, to 0.2 ml of the incubation sample an equal volume of 0.15M KOH in 85% DMSO was added to stop the reaction. Subsequently, unmetabolized B(a)P was extracted two times with 2.5 ml of hexane using maximum speed of super-mixer (Lab-Line Instruments Inc., Melrose Park, IL). Extracts were centrifuged for five minutes at 1,000 x g, 0.2 ml of water phase acidified with equal volume of 1M HCl, and, the radioactivity of the sample was counted by liquid scintillation spectrometry. The total production of B(a)P metabolites estimated either from HPLC analysis (by summation of radioactivity under the various peaks as dpm), or by the modified method of DePierre *et al.*, (1975) described above, gave comparable results. Thus, in control, unperoxidized microsomes, the amount of B(a)P metabolized was estimated as 2434 ± 241 and 2315 ± 259 pmole/mg protein *per* 10 minutes for triplicates by either method (Table 1). It should be emphasized that the summation of the radioactivity due to different metabolites of BaP, recognized as various peaks, was done after conversion of the peak counts to dpm. In this manner, variations in counting efficiency due to solvent gradient were rectified.

Data shown in Table 1 were aimed to test also the effect of 10 mM UDPGA on the overall metabolism of benzo(a)pyrene by mouse liver microsomes supplemented with NADPH. Product formation was evaluated using two different methods: (1) HPLC analysis of the metabolic products *plus* estimation of the amount bound to the microsomal macromolecules (protein), and (2) estimation of total radioactive products remaining after extraction of nonmetabolized B(a)P with hexane, according to the method of VanCantfort *et al.*, (1977). As can be seen from Table 1 the presence of UDPGA changed very little the quantity of B(a)P that was metabolized. An apparent decrease, by some 20% of the amount of radioactive derivatives of B(a)P remaining after extraction with hexane, when UDPGA was present, may be due to salting out effect of 10 mM UDPGA on some B(a)P and its lipophilic derivatives (results not shown here).

Table 1. Effect of UDPGA on overall metabolism of benzo(a)pyrene by mouse liver uninduced microsomes *in vitro*

	B(a)P Products Formation (pmol/mg prot)		
Method	Control	*plus* UDPGA	(n) P
HPLC	2434±241	2352±468	(10) n.s.
Extraction	2315±259	1866±103*	(4) 0.019

Incubation conditions: 10 min at 37°C in the medium containing 50 μM [^{14}C] B(a)P (0.585 μCi/incubation) *plus* or *minus* 10 mM UDPGA, 5 mM NADPH, 130 mM NaCl, 5.2 mM KCl, 1.3 mM KH_2PO_4, 10 mM Na_2HPO_4 (pH 7.4), 1.3 mM $MgSO_4$, 2.5 mg/ml of microsomes, final volume of 0.2 ml.

Methanolic extracts were taken for HPLC. For other details see Materials and Methods.

*Significant difference $p \leq 0.05$ by paired t-statistic. Values represent mean ± S.D., number of determinants (n) shown in parenthesis; n.s.-nonsignificant.

Another experiment checked the validity and illustrates the value of the summation of radioactivity under the peaks. It was desired to learn to what extent various metabolites of B(a)P are further metabolized to glucuronides, if the appropriate cofactor (UDPGA) is present. Knowing that the very highly polar peaks represented an unresolved mixture of glucuronides, and that the various hydroxylated metabolites of B(a)P, listed in Table 2, were potential substrates, the questions were: Do decreases in the various hydroxylated metabolites, caused by the presence of UDPGA, represent glucuronide formation? If so, do these decreases add up to the amount of polar metabolites seen under the glucuronide peak?

The effect of UDPGA on the HPLC profile of [^{14}C] benzo(a)pyrene metabolites produced by native mouse liver microsomes, with NADPH, is shown in Figure 3 and the quantitative data in Table 2. Calculation of the sum of partial decrements in each peak due to UDPGA revealed the amount of metabolites bound by glucuronidation to equal 1017 pmol/mg protein, for three parallel incubations. This value approximated the amount of metabolites detected under the glucuronide peaks, which was 1222 ± 128 pmol/mg protein for three respective parallel incubations (Table 2).

DISCUSSION

HPLC is commonly used to separate and quantitate metabolites of benzo(a)pyrene in different biosystems since 1974 (Thakker *et al.*, 1978a). Typically, in the classical procedure (Selkirk *et al.*, 1974) the benzo-(a)pyrene and its oxygenated metabolites are extracted from the incubation

Table 2. Comparison of determination of glucuronide peak directly and by calculation

	Amount of B(a)P Product (pmol/mg prot)	
Metabolites	Without UDPGA	With UDPGA
Tetrols/triols	135 ± 39	144 ± 27
Diols	170 ± 52	125 ± 61
Diones	367 ± 38	242 ± 52*
4,5-Epoxy	5 ± 10	0 ± 0
Monohydroxy	1493 ± 145	705 ± 72*
Protein bound	128 ± 28	74 ± 8*
Amount of glucuronides calculated[a]	1017	
Amount of glucuronides measured as a peak	1222 ± 128	

The incubation of microsomes with [^{14}C] B(a)P 50 μM in the presence of 5 mM NADPH were the same as in Table 1.

[a]Decrement due to UDPGA was calculated for each metabolite by subtracting from the average amounts of each metabolite formed in the absence of UDPGA that formed in its presence, and then summation of the differences gave the calculated total amount of glucuronides. Amounts of products are ± S.D. from the three determinations by HPLC of methanolic extracts. For other details see Materials and Methods.

*Significantly different from the corresponding controls $p \leq 0.05$ by paired t-statistic.

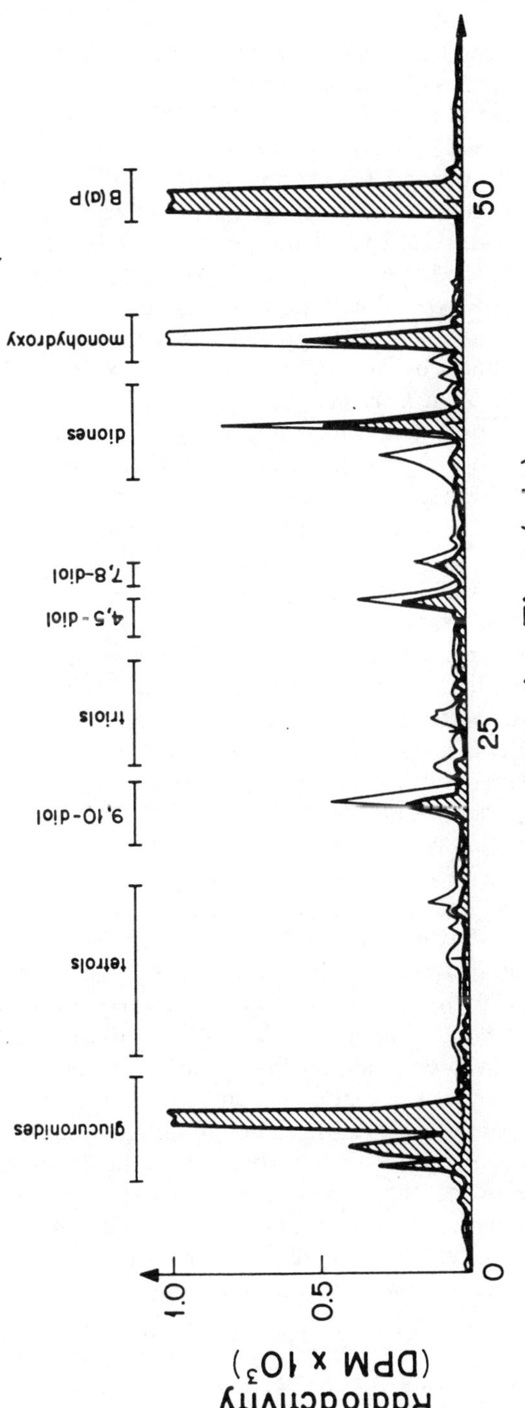

Figure 3. HPLC profiles of metabolic products of oxygenation and glucuronidation of benzo(a)pyrene by mouse liver microsomes. Native mouse liver microsomes were incubated, as described in Materials and Methods, in the presence of 5 mM NADPH and 50 μM [^{14}C] benzo(a)pyrene (0.58 μCi/incubation) without UDPGA or with 10 mM UDPGA (shaded area). For experimental details see Materials and Methods.

mixture with organic solvent, and after concentration by evaporation in a stream of N_2, injected into HPLC and analyzed. Conjugated metabolites are mostly too polar to be quantitatively extracted with ethyl acetate or other organic solvent, moreover, they are likely to partition substantially between aqueous and organic phases (Cohen *et al.*, 1976; Zaleski *et al.*, 1983). The usual methodological approach to quantitate the conjugated metabolites is hydrolysis with an appropriate enzyme (glucuronidase and/or sulphatase), and after another organic solvent extraction, analysis of the released oxygenated metabolites with HPLC (Cohen *et al.*, 1976; Selkirk *et al.*, 1980; Zaleski *et al.*, 1983). Disadvantages of this method were (1) lack of an enzyme able to hydrolyze glutathione conjugates to lipophilic compounds; (2) incomplete hydrolysis of some glucuronides and sulfates; and (3) extractability of some sulfates by organic solvents (Cohen *et al.*, 1976) as well as some glucuronides (Zaleski *et al.*, 1983).

An attempt was made to develop a direct HPLC assay for water-soluble metabolites of benzo(a)pyrene remaining in aqueous phase after extraction with ethyl acetate (Merrick and Selkirk, 1985), however, this still entails the problem of incomplete solvent extraction. Zaleski *et al.*, (1983) bypassed the solvent partitioning problem by applying aqueous acetone extracts to TLC plates for analysis of the conjugates.

Therefore, for our studies of benzo(a)pyrene oxygenation, its one-electron oxidation and glucuronidation we developed an HPLC assay which used direct methanolic extracts, as described under Methods, for separation and quantitation of water-soluble glucuronides as well as hydrophilic and hydrophobic products in a single-step analysis. To avoid extraction with hydrophobic solvents, the reaction in incubation samples was terminated with chilled methanol. Addition of 3 vol of 100% methanol to 2 vol of incubation medium apparently stopped the enzymatic reactions and resulted in the solubilization of oxygenated derivatives of benzo(a)pyrene, as they appeared in the final extract in proportions similar to that of the ethylacetate extract. Moreover, the total amount of benzo(a)pyrene metabolites measured by this method agreed with estimation by the modified method of DePierre *et al.*, (1975) as described earlier by Byczkowski and Gessner (1987a). To avoid contamination of the HPLC column with impurities or possible traces of the precipitate in direct extracts from biological samples, we connected an in-line filter after the injector, followed by a guard-pak. Frequent cleaning of the in-line filter and changing of the disposable guard-pak (every 25-50 samples) efficiently prevented the pressure build-ups and prolonged the life of the analytical column. In our experience, the application of direct methanolic extracts has proved a satisfactory method for simultaneous analysis of oxygenation and glucuronidation products of [^{14}C] benzo(a)pyrene by microsomes (Fig. 2; see also Byczkowski and Gessner 1987a; 1987b).

The problem with the analysis of [^{3}H] benzo(a)pyrene metabolites is that it gives sometimes high amounts of $^{3}H_2O$ (appearing as a first polar "flow-through" peak in the HPLC pattern, Fig. 1 and Fig. 2), which may overlap and interfere with other polar peaks that appear during one-electron oxidation of benzo(a)pyrene in the presence of added NADPH (Byczkowski and Gessner, 1987a). Also, it may disturb estimation of conjugates with glucuronate (Fig.2). Therefore, the HPLC analysis with the methanol/water solvent program was not suitable for simultaneous analysis of oxygenated and glucuronidated products of [^{3}H] benzo(a)pyrene. On the other hand, the same method gave consistent and reproducible results with [^{14}C] benzo(a)pyrene metabolized by microsomes with NADPH in the presence and absence of UDPGA (Fig. 3).

The method described here allowed, in a one-step HPLC analysis, to separate and quantitate both glucuronide-conjugated and unconjugated benzo(a)pyrene metabolites, without additional organic solvent extractions and/or enzymatic hydrolysis. It simplified the procedure and decreased time necessary to process the sample. These last features are especially important during analysis of series of many samples and controls during investigations of factors affecting activation-deactivation balance in benzo(a)pyrene metabolism.

Using a radio flow detector in series with the HPLC system, it was possible to reduce further the time of analysis, obtaining the metabolic profile of the sample within about 60 minutes, i.e., within the duration of the chromatographic run itself. Moreover, application of the computer made it possible to express the results directly in dpms, with aid of a quench-curve prepared according to the operation manual provided by the manufacturer of the flow detector (see Materials and Methods). Using floppy discs to record each run, one also has the option of reprocessing each run at a convenient time, on a suitable scale, adjusting the background subtraction and the minimal integrated peak area, as well as to store and file the results in a very efficient way.

Usage of liquid scintillation counting in the radio flow detector with appropriate selection of the cell size and the flow rate, gave a method of good sensitivity for studies of metabolite profiles. Counting efficiency on the order of 30% for ^{3}H and 70% for ^{14}C, resembles that of conventional counting in a stationary liquid scintillation spectrometer. Suitability of the described method to separate and quantitate the oxygenation and oxidation products of benzo(a)pyrene along with their glucuronides has been confirmed in several studies involving enzymatic and nonenzymatic systems (Byczkowski and Gessner, 1986; 1987a, 1987b, 1987c).

ACKNOWLEDGEMENTS

This research was supported in part by U.S.P.H.S. Grant CA- 24127 from National Cancer Institute. Thanks are due to the Chemical and Physical Carcinogenesis Branch of the N.C.I. for benzo(a)pyrene derivatives.

ABBREVIATIONS USED

HPLC = high performance (high pressure) liquid chromatography
NADPH = β-nicotinamide adenine dinucleotide phosphate reduced form
UDPGA = uridine 5′-diphosphoglucuronic acid
B(a)P = benzo(a)pyrene
7,8,9,10-tetrol = 7,8,9,10-tetrahydroxy-7,8,9,10-tetrahydrobenzo(a)pyrene
9,10-diol = 9,10-dihydrobenzo(a)pyrene-9,10-diol
7,8,9,-triol = 7,8,9-trihydroxy-7,8,9,10-tetrahydrobenzo(a)pyrene
4,5-diol = 4,5-dihydrobenzo(a)pyrene-4,5-diol
7,8-diol = 7,8-dihydrobenzo(a)pyrene-7,8-diol
1,6-dione = benzo(a)pyrene-1,6-dione
3,6-dione = benzo(a)pyrene-3,6-dione
6,12-dione = benzo(a)pyrene-6,12-dione
4,5-epoxy = 4,5-dihydrobenzo(a)pyrene-4,5-epoxide
9-monohydroxy = 9-hydroxybenzo(a)pyrene
3-monohydroxy = 3-hydroxybenzo(a)pyrene
Fe^{3+}·FeEDTA = equimolar mixture of $Fe_2(SO_4)_3$ with $Fe(NH_4)_2$ $(SO_4)_2$ and ethylenediamine-tetraacetic acid
7,8-diol 9,10-epoxide-2 = (±)-7β, 8α-dihydroxy-9α, 10α-epoxy-7,8,9,10-tetrahydrobenzo(a)pyrene-2.

REFERENCES

Ashurst, S.W., Mehta, R. and Cohen, G.M. (1979). Importance of conjugation reactions in determining the qualitative nature of polycyclic aromatic hydrocarbon-DNA interactions. *Medical Biol.* 57, 313-320.

Autrup, H. (1979). Separation of water-soluble metabolites of benzo(a)pyrene formed by cultured human colon. *Biochem. Pharmacol.* 28, 1727-1730.

Autrup, H., Harris, C.C., Stoner, G.D., Selkirk, J.L., Schafer, P.W. and Trump, B.F. (1978). Metabolism of [^{3}H] benzo(a)pyrene by cultured human bronchus and cultured human pulmonary alveolar macrophages. *Lab. Invest.* 38, 217-224.

Ball, L.M., Plummer, J.L., Smith, B.R. and Bend, J.R. (1979). Benzo(a)pyrene oxidation, conjugation and disposition in the isolated perfused rabbit lung: role of the glutathione S-transferases. *Medical Biol.* 57, 298-305.

Bansal, S.K., Zaleski, J. and Gessner, T. (1981). Glucuronidation of oxygenated benzo(a)pyrene derivatives by UDP-glucuronyltransferase of nuclear envelope. *Biochem. Biophys. Res. Comm.* 98, 131-139.

Bend, J.R., Smith, B.R., Ball, L.M., Plummer, J.L., Wolfe, C.R., Philpot, R.M., Devereux, T.R. and Fouts, J.R. (1981). Metabolism of benzo(a)pyrene and benzo(a)pyrene 4,5-oxide in rabbit lung. *Adv. Exp. Med. Biol.* 136, 541-554.

Bond, J.A. (1983). Some biotransformation enzymes responsible for polycyclic aromatic hydrocarbon metabolism in rat nasal turbinates: Effects on enzyme activities of *in vitro* modifiers and intraperitoneals and inhalation exposure of rats to inducing agents. *Cancer Res.* 43, 4805-4811.

Byczkowski, J.Z. and Gessner, T. (1986). Modification of microsomal metabolism of benzo(a)pyrene (B(a)P) by superoxide generated *in vitro. Fed. Proc.* 45, 3146.

Byczkowski, J.Z. and Gessner, T. (1987a). Action of xanthine-xanthine oxidase system on microsomal benzo(a)pyrene metabolism *in vitro. Gen. Pharmacol.* 18, 385-395.

Byczkowski, J.Z. and Gessner, T. (1987b). Effects of superoxide generated *in vitro* on glucuronidation of benzo(a)pyrene metabolites by mouse liver microsomes. *Int. J. Biochem.*, 19, 531-537.

Byczkowski, J.Z. and Gessner, T. (1987c). Asbestos-catalyzed oxidation of benzo(a)pyrene by superoxide-peroxidized microsomes. *Bull. Environm. Contam. Toxicol.*, 39,312-317.

Cohen, G.M., Haws, S.M., Moore, B.P. and Bridges, J.W. (1976). Benzo(a)pyrene 3yl hydrogen sulphate, a major ethyl acetate-extractable metabolite of benzo(a)pyrene in human, hamster and rat lung cultures. *Biochem. Pharmacol.* 25, 2561-2570.

Cook, J.W., Ludwiczak, R.S. and Schoental, R. (1950). Polycyclic aromatic hydrocarbons. Part XXXVI. Synthesis of the metabolic oxidation products of 3:4-benzopyrene. *J. Chem. Soc.* p. 11, 1112-1121.

Cunningham, M.L. and Lokesh, B.R. (1983). Superoxide anion generated by potassium superoxide is cytotoxic and mutagenic to Chinese hamster ovary cells. *Mutation Res.* 121, 299-304.

Daniel, P.M., Pratt, O.E. and Prichard, M.M.L. (1967). Metabolism of labeled carcinogenic hydrocarbons in rats. *Nature* 215, 1142-1146.

DePierre, J.W., Moron, M.S., Johannesen, K.A.M. and Ernster, L. (1975). A reliable, sensitive and convenient radioactive assay for benzopyrene monooxygenase. *Analyt. Biochem.* 63, 470-484.

Dix, T.A. and Marnett, L.J. (1984). Detection of the metabolism of polycyclic aromatic hydrocarbon derivatives to ultimate carcinogens during lipid peroxidation. *Methods Enzymol.* 105, 347-352.

Dix, T.A., Fontana, R., Panthani, A. and Marnett, L.J. (1985). Hematin catalyzed epoxidation of 7,8-dihydroxy-7,8- dihydrobenzo(a)pyrene by polyunsaturated fatty acid hydroperoxides. *J. Biol. Chem.* 260, 5358-5365.

Falk, H.L., Kotin, P., Lee, S.S. and Nathan, A. (1962). Intermediary metabolism of benzo(a)pyrene in the rat. *J. Nat. Cancer Inst.* 28, 699-724.

Gelboin, H.V. (1980). Benzo(a)pyrene metabolism, activation, and carcinogenesis: role and regulation of mixed-function oxidases and related enzymes. *Physiol. Rev.* 60, 1107-1166.

Gelboin, H.V., Okuda, T., Selkirk, J., Nemoto, N., Yang, S.K., Wiebel, F.J., Whitlock, J.P., Rapp, H.J. and Bast, Jr., R.C. (1976). Benzo(a)pyrene metabolism: Enzymatic and liquid chromatographic analysis and application to human liver, lymphocytes and monocytes. In: *Screening Tests in Chemical Carcinogenesis*, R. Montesano, H. Bartsch and L. Tomatis (eds.). IARC Scientific Publications, No. 12, Lyon, pp. 225-254.

Haake, J.M., Merrill, J.C. and Safe, S. (1985). The *in vitro* metabolism of benzo(a)pyrene by polychlorinated and polybrominated biphenyl induced rat hepatic microsomal monooxygenases. *Can. J. Physiol. Pharmacol.* 63, 1096-1100.

Holder, G., Yahi, H., Dansette, P., Jerina, D.M., Levin, W., Lu, A.Y.M. and Conney, A.H. (1974). Effects of inducers and epoxide hydrase on the metabolism of benzo(a)pyrene by liver microsomes and a reconstituted system: analysis by high pressure liquid chromatography. *Proc. Nat. Acad. Sci. USA* 71, 4356-4360.

Holder, G.M., Yagi, H., Jerina, D.M., Levin, W., Lu, A.Y.H. and Conney, A.H. (1975). Metabolism of benzo(a)pyrene. Effect of substrate concentration and 3-methylcholanthrene pretreatment on hepatic metabolism by microsomes from rats and mice. *Arch. Biochem. Biochys.* 170, 557-566.

Jones, C.A., Moore, B.P., Cohen, G.M., Fry, J.R. and Bridges, J.W. (1978). Studies on the metabolism and excretion of benzo(a)pyrene in isolated adult rat hepatocytes. *Biochem. Pharmacol.* 27, 693-702.

Keller, G.M., Turner, C.R., and Jefcoate, C.R. (1982). Kinetic determinants of benzo(a)pyrene metabolism to dihydrodiol epoxides by 3-methylcholanthrene-induced rat liver microsomes. *Mol. Pharmacol.* 22, 451-458.

Lam, K.T., Wattenberg, L.W. (1977). Effects of butylated hydroxyanisole on the metabolism of benzo(a)pyrene by mouse liver microsomes. *J. Natl. Cancer Inst.* 58, 413-417.

Lesko, S.A., Lorentzen, R.J. and Ts'o, P.O.P. (1980). Role of superoxide in deoxyribonucleic acid strand scission. *Biochemistry* 19, 3023-3028.

Marshall, M.V., Gonzalez, M.A., McLemove, T.L., Busbee, D.L., Wray, N.P. and Griffin, A.C. (1980). Reversed-phase separation of benzo(a)pryene metabolites by thin-layer chromatography. *J. Chromatogr.* 197, 217-225.

Marshall, M.V., McLemore, T.L., Martin, R.R., Jenkins, W.T., Snodgrass, D.R., Carson, M.A., Arnott, M.S., Wray, N.P. and Griffin, A.C. (1979). Patterns of benzo(a)pyrene metabolism in normal human pulmonary alveolar macrophages. *Cancer Lett.* 8, 103-109.

Merrick, B.A. and Selkirk, J.K. (1985). HPLC of benzo(a)pyrene glucuronide, sulfate and glutathione conjugates and water-solube metabolites from hamster embryo fibroblasts. *Carcinogenesis* 6, 1303-1307.

Moore, C.J. and Gould, M.N. (1984). Metabolism of benzo(a)pyrene by cultured human hepatocytes from multiple donors. *Carcinogenesis* 5, 1577-1582.

Murphy, S.E. and Hecht, S.S. (1985). Dual-label high-performance liquid chromatographic assay for fentomole levels of benzo(a)pyrene metabolites. *Analyt. Biochem.* 146, 442-447.

Nishimoto, M. and Varnasi, U. (1985). Benzo(a)pyrene metabolism and DNA adduct formation mediated by English sole liver enzymes. *Biochem. Pharmacol.* 34, 263-268.

Oesch, F., Bentley, P., Golan, M. and Stasiecki, P. (1985). Metabolism of benzo(a)pyrene by subcellular fractions of rat liver: evidence for similar patterns of cytochrome P-450 in rough and smooth endoplasmic reticulum but not in nuclei and plasma membrane. *Cancer Res.* 45, 4838-4843.

Pitts, Jr., J.N., Lokensgard, D.M., Ripley, P.S., VanCouwenberghe, K.A., VanVaeck, L., Shaffer, S.D., Thill, A.J. and Belser, Jr., W.L. (1980). ''Atmospheric'' epoxidation of benzo(a)pyrene by ozone: Formation of the metabolite benzo(a)pyrene-4,5-oxide. *Science* 210, 1347-1349.

Raha, C.R. (1972). Metabolism of benzo(a)pyrene at 4,5-position. *Indian J. Biochem. Biophys.* 9, 105-110.

Reed, G.A. and Marnett, L.J. (1982). Metabolism and activation of 7,8-dihydrobenzo(a)pyrene during prostaglandin biosynthesis. *J. Biol. Chem.* 257, 11368-11376.

Sadowski, I.J., Wright, J.A. and Israels, L.G. (1985). A permeabilized cell system for studying regulation of aryl hydrocarbon hydroxylase: NADPH as rate limiting factor in benzo(a)pyrene metabolism. *Int. J. Biochem.* 17, 1023-1025.

Selkirk, J.K. (1980). Comparison of epoxide and free-radical mechanisms for activation of benzo(a)pyrene by Sprague-Dawley rat liver microsomes. *J. Natl. Cancer Inst.* 64, 771-774.

Selkirk, J.K., Cohen, G.M. and MacLeod, M.C. (1980). Glucuronic acid conjugation in the metabolism of chemical carcinogens by rodent cells. *Arch. Toxicol. Suppl.* 3, 171-178.

Selkirk, J.K., Croy, R.G., Roller, P.P. and Gelboin, H.V. (1974). High-pressure liquid chromatographic analysis of benzo(a)pyrene metabolism and covalent binding and the mechanism of action of 7,8-benzoflavone and 1,2-epoxy-3,3,3-trichloropropane. *Cancer Res.* 34, 3474-3480.

Selkirk, J.K., Croy, R.G. and Gelboin, H.V. (1974a). Benzo(a)pyrene metabolites efficient and rapid separation by high-pressure liquid chromatography. *Science* 184, 169-171.

Selkirk, J.K., Croy, R.G. and Gelboin, H.V. (1975). Isolation by high pressure liquid chromatography and characterization of benzo(a)pyrene-4,5-epoxide as a metabolite of benzo(a)pyrene. *Arch. Biochem. Biophys.* 168, 322-326.

Sims, P. (1967). The metabolism of benzo(a)pyrene by rat-liver homogenates. *Biochem. Pharmacol.* 16, 613-618.

Sims, P., Grover, P.L., Swaisland, A., Pal, K. and Hemer, A. (1974). Metabolic activation of benzo(a)pyrene proceeds by a diol-epoxide. *Nature* 252, 326-328.

Smith, B.R. and Bend, J.R. (1979). Metabolism and toxicity of benzo(a)pyrene-4,5-oxide in the isolated perfused rat liver. *Toxicol. Appl. Pharmacol.* 49, 313-321.

Smith, B.R., Philpot, R.M. and Bend, J.R. (1978). Metabolism of benzo(a)pyrene by the isolated perfused rabbit lung. *Drug Metab. Dispos.* 6, 425-431.

Smith, B.R., Plummer, J.L., Ball, L.M. and Bend, J.R. (1980). Characterization of pulmonary arene oxide biotransformation using the perfused rabbit lung. *Cancer Res.* 40, 101-106.

Sweeny, D.J. and Reinke, L.A. (1986). Production and release of benzo(a)pyrene phenols from the perfused rat liver. *Fed. Proc.* 45, 1679.

Takahasi, G., Kinoshita, K., Hashimoto, K. and Yasuhira, K. (1979). Identification of benzo(a)pyrene metabolites by gas-chromatograph-mass spectrophotometer. *Cancer Res.* 39, 1814-1818.

Thakker, D.R., Yagi, H., Levin, W., Wood, A.W., Conney, A.H. and Jerina, D.M. (1985). Polycyclic aromatic hydrocarbons: Metabolic activation to ultimate carcinogens. In: *Bioactivation of Foreign Compounds,* Academic Press Inc., New York, pp. 177-242.

Thakker, D.R., Yagi, H., Lehr, R.E., Levin, W., Buening, M., Lu, A.Y.M., Chang, R.L., Wood, A.W., Conney, A.M. and Jerina, D.M. (1978). Metabolism of trans-9,10-dihydroxy-9,10-dihydrobenzo(a)pyrene occurs primarily by arylhydroxylation rather than formation of a diol epoxide. *Mol. Pharmacol.* 14, 502-513.

Thakker, D.R., Yagi, H. and Larina, D.M. (1978a). Analysis of polycyclic aromatic hydrocarbons and their metabolites by high-pressure liquid chromatography. *Methods Enzymol.* 52, 279-296.

Thakker, D.R., Yagi, H., Akagi, H., Koreeda, M., Lu, A.Y.H., Levin, W., Wood, W., Cooney, A.H. and Jerina, D.M. (1977). Metabolism of benzo(a)pyrene. VI. Stereoselective metabolism of benzo(a)pyrene and benzo(a)pyrene 7,8-dihydrodiol to diol epoxides. *Chem. Biol. Interact.* 16, 281-300.

VanCantfort, J., Degraeve, J. and Gielen, J.E. (1977). Radioactive assay for aryl hydrocarbon hydroxylase. Improved method and biological importance. *Biochem. Biophys. Res. Comm.* 79, 505-512.

Vaught, J.B., Gurtoo, H.L., Paigen, B., Minowada, J. and Sartori, P. (1978). Comparison of benzo(a)pyrene metabolism by human peripheral blood lymphocytes and monocytes. *Cancer Lett.* 5, 261-268.

Warshawsky, D., Bingham, E. and Niemeier, R.W. (1984). The effects of cocarcinogen, ferric oxide, on the metabolism of benzo(a)pyrene in the isolated perfused lung. *J. Toxicol. Environ. Health* 14, 191-209.

Warshawsky, D., Bingham, E. and Niemeier, R.W. (1983). Influence of airborne particulate on the metabolism of benzo(a)pyrene in the isolated perfused lung. *J. Toxicol. Environ. Health* 11, 503-517.

Warshawsky, D., Niemeier, R. and Bingham, E. (1981). Influence of sulfur dioxide on metabolism and distribution of benzo(a)pyrene in isolated perfused rabbit lung. *J. Toxicol. Environ. Health* 7, 1001-1024.

Waterfall, J.F. and Sims, P. (1972). Epoxy derivatives of aromatic polycyclic hydrocarbons. The preparation and metabolism of epoxides related to benzo(a)pyrene and to 7,8- and 9,10- dihydrobenzo(a)pyrene. *Biochem. J.* 128, 265-277.

Weems, H.B. and Yang, S.K. (1982). Resolution of optical isomers of chiral high-performance liquid chromatography: separation of dihydrodiols and tetrahydrodiols of benzo(a)pyrene and benzo(a)anthracene. *Analyt. Biochem.* 125, 156-161.

Weigert, F. and Mottram, J.C. (1946). The biochemistry of benzopyrene. I. and II. *Cancer Res.* 6, 97-120.

Yang, S.K., Roller, P.P. and Gelboin, H.V. (1977). Enzymatic mechanism of benzo(a)pyrene conversion to phenols and diols and an improved high-pressure liquid chromatographic separation of benzo(a)pyrene derivatives. *Biochemistry* 16, 3680-3687.

Yang, S.K., Selkirk, J.K., Plotkin, E.V. and Gelboin, H.V. (1975). Kinetic analysis of the metabolism of benzo(a)pyrene to phenols, dihydrodiols, and quinones by high-pressure chromatography compared to analysis by aryl hydrocarbon hydroxylase assay, and the effect of enzyme induction. *Cancer Res.* 35, 3642-3650.

Zaleski, J., Bansal, S.K. and Gessner, T. (1983). Formation of glucuronide sulphate and glutathione conjugates of benzo(a)pyrene metabolites in hepatocytes isolated from inbred strains of mice. *Carcinogenesis* 4, 1359-1366.

Progress in HPLC, Vol. 3, pp. 167-189
Parvez et al. (Eds)

Radioisotope detectors for investigations on phenolic biosynthesis

JAMES A. SAUNDERS
Germplasm Quality & Enhancement Lab, USDA, Beltsville Agriculture Research Center Beltsville, MD USA

and JOSEPH OLECHNO
Dionex Corporation, 1228 Titan Way, Sunnyvale, CA 94088 USA

INTRODUCTION AND OVERVIEW

Secondary metabolites exist throughout the plant kingdom as an important, yet poorly understood class of chemical compounds (Bell and Charlwooded, 1980; Bell, 1981). They are grouped in a variety of ways: by their chemical structure (the flavonoids, steroids, stilbenes, xanthones); by their chemical characteristics (alkaloids, phenolics, quinones); by their functional components (glycosides, lipids); and by their physiological function (as growth regulators, sunscreens, allelochemicals). There are overlaps between these categories, for example, tomatine is both an alkaloid and a glycoside, and rutin is a flavonoid, a phenolic, and a glycoside. However, these categories enable the researcher to study major classes of compounds with similar characteristics.

Phenolic compounds are one of the largest and most ubiquitous group of heterogeneous secondary metabolites and occur in all higher plants as well as fungi and mosses (Harborne, 1980). The term "phenolics" is a generic phrase that includes a wide range of chemical compounds, all of which possess an aromatic ring with at least one hydroxyl substituent. Thousands of phenolics are known and they form a number of large groups having structural similarities. The largest of these groups is the flavonoids that are responsible for many of the red, blue, and yellow pigments as well as many colorless compounds in plants. These compounds, reviewed by Harborne (1967) Harborne et al., (1982), are closely related by structure. They are functionally involved in a variety of metabolic processes, including plant defenses from insects and pathogens, pollen, and nectar guides, hormone regulation, protection from UV irradiation, and others.

This chapter will discuss some of the early steps in the biosynthesis of phenolics and the reliability of the enzymatic assays that have been used on three key enzymes involved in the biosynthesis of these secondary natural products (Fig. 1). The enzymes phenylalanine ammonia lyase (PAL), tyrosine ammonia lyase (TAL), and cinnamic acid hydroxylase (CAH) were assayed using radioisotopic substrates and the reaction products separated by means of high pressure liquid chromatography (HPLC) equipped with a radioisotope detector. Using these procedures we have attempted to clarify conflicting reports in the literature concerning these enzymes and their reaction products.

INTRODUCTION TO PHENOLIC BIOSYNTHESIS

The amino acids tryptophan, phenylalanine, and tyrosine (itself a phenolic) give rise to a large number of secondary metabolites in plants, including a vast array of alkaloids and non-nitrogenous phenolics. The biosyntheses of phenylalanine and tyrosine have been reviewed by Davis (1958).

At the turn of the century, Collie (1893, 1907) suggested that aromatic compounds of plants might arise from the condensation of acetate subunits. While structural analysis supported this view, it remained for Birch et al. (1955, 1958a, 1958b) to confirm that acetate was indeed a precursor of some phenolics in plants. Gatenbeck and Mosbach (1959), using double labeled radioisotopic precursors, showed that acetate molecules are linked head to tail and cyclized to produce some aromatic phenolics.

Although these "acetogenins" account for many of the naturally occurring phenolics, particularly the quinones, there are other families of phenolics, such as protocatechuic and gallic acids in fungi, that are derived from the shikimic acid pathway or from the metabolism of the aromatic amino acid, tryptophan. Yet still others are derived through the deamination

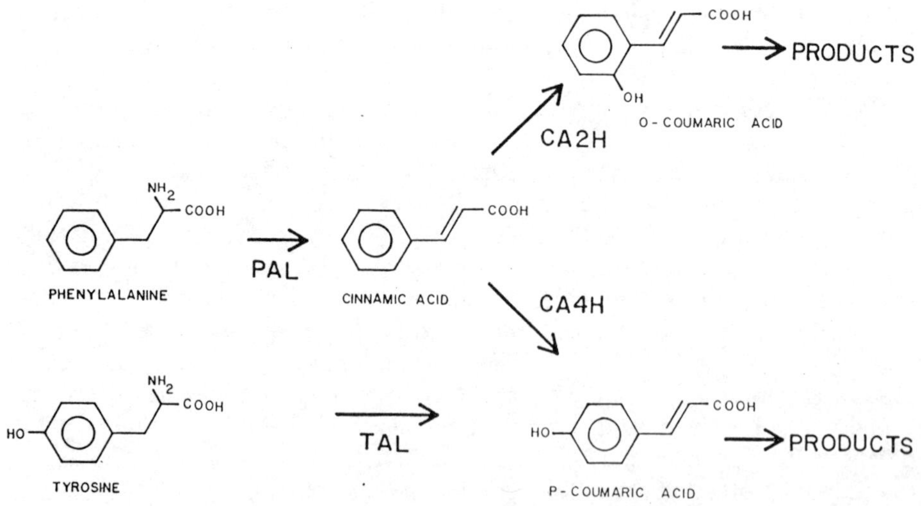

Figure 1. "Hub" reactions of the phenylpropanoid pathway showing the conversion of phenylalanine and tyrosine to cinnamic and p-coumaric acid, respectively.

and subsequent hydroxylation of the amino acids, phenylalanine, and tyrosine. This last group of phenolics is the phenylpropanoid group and the resultant hydroxylated cinnamic acids constitute a large range of varied structures when condensed with malonyl CoA (Kreuzler and Hahlbrook, 1972, 1975; Swain et al., 1979). These hydroxylated cinnamic acids may be reduced to alcohols and aldehydes, or polymerized to lignans (Gross et al., 1973; Higuchi et al., 1977) to produce a wide array of phenolics.

HPLC SEPARATION TECHNIQUES

HPLC separation techniques are in widespread use for thousands of organic compounds derived from biological sources. The isolation of secondary metabolites by HPLC procedures, especially alkaloids and phenolics, have been investigated by the researchers in various laboratories (Krstulovic and Brown, 1982; Sisson and Saunders, 1982; Court, 1977).

While separation with HPLC is a rapid technique that has been successfully used with a great many compounds, many of the standard methods of detection, UV-visible absorbance, refractive index, and electro-analytical procedures are often relatively insensitive to certain compounds, or require time-consuming derivatization steps that are cumbersome or unreliable. Radioisotopes have been used to determine the biochemical synthesis and metabolism of many of these secondary metabolites, and the use of HPLC and radioisotope detection, in tandem, represents a powerful analytical tool to supplement other separation procedures that are plagued by limited resolution, prolonged analysis time, and potential misidentifications.

Using an isocratic HPLC system with a reverse phase column we were able to separate the trans-isomers of para, ortho, and meta-coumaric acids (p-,o- and m-coumaric acid), cinnamic acid, caffeic acid, and coumarin within a 25 minute chromatographic profile (Fig. 2). These compounds represent several of the key intermediates in the production of many phenolics and coumarins.

PHENYLALANINE AMMONIA LYASE

Phenylalanine ammonia lyase (PAL) is the first committed enzyme of the phenylpropanoid pathway. It catalyzes the conversion of phenylalanine to trans-cinnamic acid (Fig. 1) through the loss of the alpha-amino group and the pro-3S hydrogen (Hanson and Havir, 1972; Bartl et al., 1977). In some plants, tyrosine may act as a substrate with the production of trans-p-coumaric acid, although tyrosine is not a suitable substrate for PAL extracted from all sources.

A variety of other amino acid ammonia lyases have been reported, but they appear to be fundamentally different from PAL with the possible exception of histidine ammonia lyase, which may have a similar active site (Givot et al., 1969; Hanson and Havir, 1972).

PAL appears to be ubiquitous, found in the families of all higher plants, in algae and in many fungi (Camm and Towers, 1973b; Hanson and Havir,

1981). In general, the enzyme appears to be a tightly bound tetramer (Havir and Hanson, 1973), the subunits of which can only be separated by high concentrations of guanidinium chloride, urea or vigorous SDS-mercaptoethanol treatments. The molecular weight of each PAL monomer is between 83-90 Kda (Camm and Towers, 1977). The subunits, if not identical are very similar; however, the enzyme displays a negative co-operativity with respect to substrate binding. There is a strong likelihood that the enzyme exists as a glycoprotein, and that minor variations of its properties may be due to the extent of glycosylation (Havir, 1979). Binding studies with nucleophilic reagents seem to indicate that there are only two active sites, and that they may involve a dehydroalanine (Hanson and Havir, 1969; 1981). The pH optimum for PAL is about 8.8 (Havir et al., 1971).

PAL activity is found in the cytosol as well as associated with a variety of membranes and organelles including plastids, mitochondria and microbodies (Saunders and McClure, 1975; Alibert et al., 1977; McClure, 1979; Ranjeva et al., 1979). The occurrence of PAL activity in several different organelles may indicate the presence of more than one enzyme, for example, spinach appears to have at least three different activities. The extrachloroplastic cytosol has about 75% of the total PAL activity, while two other enzymes seem to account for the PAL activity of the chloroplast, one of those being regulated by thioredoxin (Nishizawa et al., 1979). Oak also appears to have at least two PAL enzymes associated with different organelles (Alibert et al., 1972a; 1972b). As with other enzymes involved in

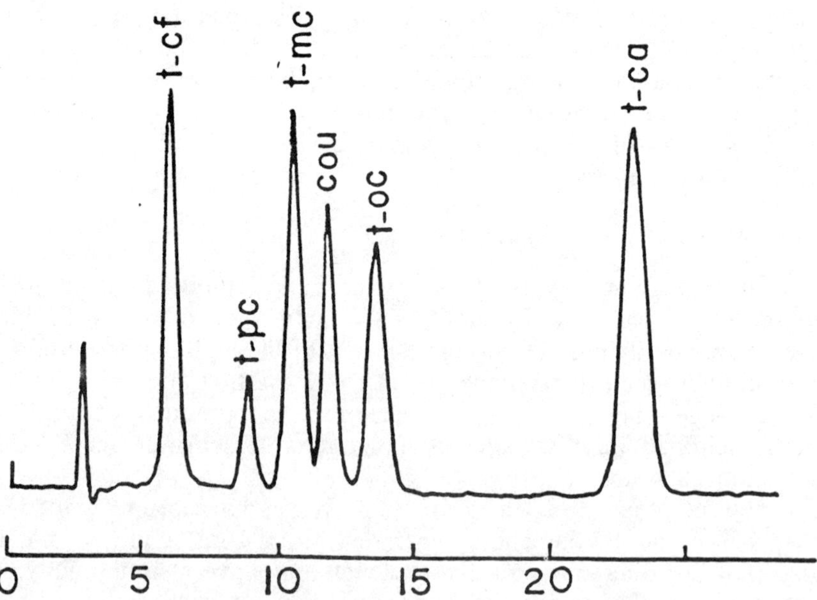

Figure 2. HPLC tracing of the trans-isomers of cinnamic acid, ortho-, meta-, and para-coumaric acid, caffeic acid, and coumarin.

the production of secondary metabolites, some PAL activities are constitutive while other PAL activities are inducible.

PAL was reported as a soluble enzyme isolated from barley shoots by Koukol and Conn (1961), although another earlier report indicated that some PAL activities may be membrane bound (Neish, 1961). Because of the importance of PAL in the biosynthesis of secondary metabolites, numerous enzymatic assays have been developed with detection of the product based upon spectrophotometric and radioisotopic techniques. Many of these assays have been criticized due to side reactions, interfering substances in the crude homogenates, or errors in the assay itself (Saunders and McClure, 1974; Poulton et al., 1980). To clarify these concerns we examined PAL activity by monitoring the conversion of radioisotopically labeled carbon 14 (^{14}C) phenylalanine to cinnamic acid by an HPLC system equipped with a Radiomatic radioisotope detector*. Using HPLC radioisotope techniques in the analysis of PAL we were able to quantitate the conversion of 1.5×10^{-12} moles of phenylalanine to cinnamic acid in our dilute assay system. The UV absorbance of cinnamic acid at that concentration would have been far below the sensitivity levels of UV detectors.

PAL activity was found in all four plants tested in this study, *Melilotus alba, Hydrangea, Heirochloe odorata,* and *Petunia*. ^{14}C Phenylalanine was converted to cinnamic acid as the sole product without the addition of exogenous cofactors or further metabolism. Figure 3 shows a HPLC tracing of an aliquot of the PAL reaction mixture of *Melilotus* taken at the start and at the end of the incubation periods. It is clear that the radioisotope detector (top tracing) can easily detect very low levels of trans-cinnamic acid (t-ca) while there is no appreciable change in the UV absorbance of the cinnamic acid at 254 nm (bottom tracing). It is also clear that additional metabolites of cinnamic acid would be detected if they were resolved chromatographically from either phenylalanine or the cinnamic acid.

TRYOSINE AMMONIA LYASE

Phenylalanine and tyrosine represent key substrates for the regulatory enzymes PAL and TAL. In some plant families, noticeably the grasses, there appears to be a tyrosine ammonia lyase that directly deaminates tyrosine to produce p-coumaric acid (Guerra et al., 1985). In most studies that have reported TAL activity in plant extracts, PAL activity is also present and the existence of separate enzymes, which are specific to tyrosine and not phenylalanine, has been questioned. Recently, however, TAL activity has been reported which is distinctly separable from PAL activity (Beaudoin-Eagan and Thorpe, 1985).

Although there were no reports of TAL activity in the four plants chosen for this study, we were interested in the applicability of the HPLC system

*Mention of a trade name or proprietary product does not constitute a guarantee of the product by these authors or their affiliated organizations nor does it imply approval to the exclusion of other products, which may also be suitable.

used for PAL for the determination of TAL. TAL activity was found in extracts from *Hydrangea* (Fig. 4). Although the chromatographic system used for the PAL assay did not fully resolve radioactive p-coumaric acid from the ^{14}C labeled tyrosine, a substantial amount of p-coumaric acid was quantitatively detected by the radioisotope detector. In addition, the HPLC system was able to detect an unkown radioactive contaminant that was present in the commercially labeled tyrosine. Although TAL activity was found in *Hydrangea*, no attempt was made to differentiate between a unique TAL enzyme and nonspecific PAL activity that could also yield the same results.

CINNAMIC ACID HYDOXYLASE

Russel and Conn (1967) and Russel (1971) first reported the isolation of cinnamic acid-4-hydroxylase (CA4H) in the microsomal fraction of *Pisum* seedlings that had been exposed to a short period of white light. Similarly, Stafford (1969) showed CA4H to be present in microsomal preparations of young *Sorghum* seedlings. This enzyme appears to be a P-450 type enzyme; it prefers NADPH as its external reductant, is inhibited by CO, is reversibly inhibited by its end product, and works best at a pH of 7-8. In *Pisum* it

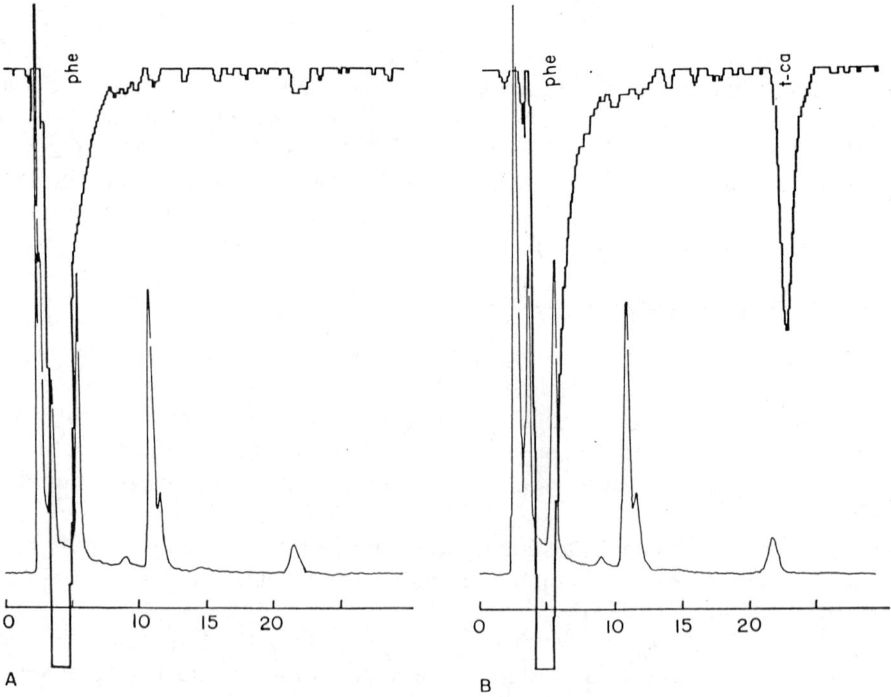

Figure 3. HPLC radioisotope tracing of a PAL assay reaction mixture. Lower tracing represents UV absorbance at 254 nm while upper tracing represents the inverted radioisotope detector signal. A: start of assay. B: two hours of incubation.

appears to work with an "NIH shift", a proton migration, away from the site of aryl hydroxylation (Ellis and Amrhein, 1971; Zahringer et al., 1978). Similar enzymes of microsomal fractions have been reported in *Brassica* (Hill and Rhodes, 1975), *Cucumis* (Billet and Smith, 1980), *Glycine* (Zahringer et al., 1978), *Petroselinum* (Buche and Sanderman, 1973), *Petunia* (Ranjeva et al., 1977), *Quercus* (Alibert et al., 1972a; 1972b), and *Solanum* (Camm and Towers, 1973a; Matthews et al., 1982).

Cinnamic acid-4-hydroxylase (CA4H), which hydroxylates cinnamic acid in the para position, has also been found as a membrane bound enzyme in the thylakoids of *Duniellia* and the membranes of some fungi (Czichi and Kindl, 1977). Although a soluble CA4H was reported by Nair and Vining (1965) that was extracted from an acetone powder of spinach, this work has not been confirmed. Benveniste et al. (1978) have concluded that the injury induced CA4H of Jerusalem artichoke tubers is associated with the endoplasmic reticulum, and Rich and Lamb (1977) have reported that the CA4H of sliced potato tubers is associated with smooth membrane vesicles that contain cytochrome P-450. Saunders et al. (1977), in agreement with Benveniste, report that the CA4H of *Sorghum* is located on the endoplasmic reticulum.

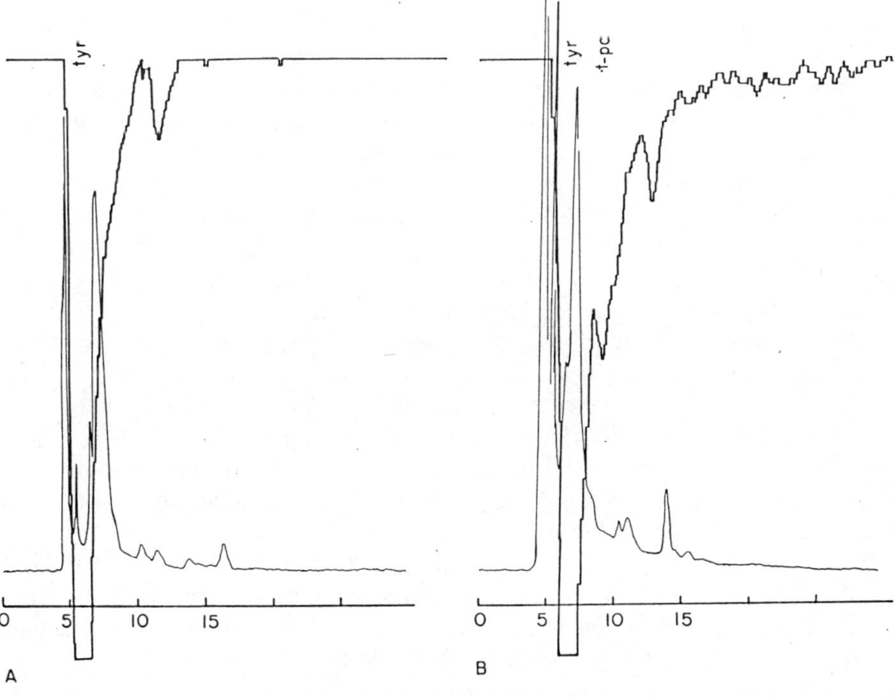

Figure 4. HPLC radioisotope tracing of a TAL assay mixture. Lower tracing represents UV absorbance at 254 nm while upper tracing represents the inverted radioisotope detector signal. A: start of assay. B: two hours of incubation.

In light of the generally accepted view that PAL, the first enzyme of the phenylpropanoid biosynthetic pathway, is a soluble enzyme, it is interesting to speculate about the association of PAL with CA4H and the other enzymes in the phenylpropanoid pathway. Stafford (1974) has suggested that these enzymes may exist in multi-enzyme complexes that are attached to membranes within the cytoplasm or are enclosed within organelles. Since PAL has been reported to occur in several different organelles, including glyoxysomes and chloroplasts, these enzyme complexes may be associated with several different organelles. This is substantiated by the observation that flavonoids are associated with the chloroplasts as well as the vacuoles (Saunders and McClure, 1976a, 1976b; Hrazdina et al., 1978; Moskowitz and Hrazdina, 1981). These complexes, like CA4H itself, may be either constitutive as in *Sorghum* or induced through injury as in *Solanum*.

The formation of p-coumaric acid from cinnamic acid has been studied in a variety of systems showing that the enzyme can oxidize cinnamic acid directly. Cinnamic acid may also be oxidized after conjugation with an activating group. p-Coumaric acid itself may be directly oxidized to caffeic, ferulic or sinapic acid (Sato, 1966) or may require activation before oxidation. There are several possible activating groups that have been examined for p-coumaric acid, including conjugates of glucose, shikimic acid, quinic acid, choline, and coenzyme A (Rhodes and Wooltorton, 1976; Zenk, 1978; Levy and Zucker, 1960; Kojima and Uritani, 1972). Once the p-coumaric acid moiety has been incorporated into a structure there may be further oxidation. Thus, it appears that modifications of the p-coumaric basic skeleton may be accomplished at the level of the free acid or after the cinnamic acid has been subsequently incorporated into various components.

The formation of coumarin requires the oxidation of cinnamic acid in the 2- or ortho-position (Fig. 5). The biological significance of coumarins in plants is not well understood although they have effects on plant growth (Staleknecht, 1972), respiration (Albu et al., 1969), and oxidative phosphorylation (Ultizur and Pollakoff-Mayber, 1963).

Whereas cinnamic acid-4-hydroxylase is widespread in plant systems, cinnamic acid-2-hydroxylase (CA2H) is considerably less prevalent. Gestetner and Conn (1974) and Ranjeva et al. (1977) reported this enzyme in the chloroplasts of *Melilotus* and *Petunia*, and Brown et al. (1960) have reported it in the monocot *Heirochloe*. Presumably the enzyme must be present in all plants that can produce coumarin or ''bound coumarin'', the glucoside of coumarinic acid (Fig. 5).

Ellis and Amrhein (1971) have reported that the CA2H enzyme, like CA4H, produces an ''NIH shift'' and retains tritium label originally in the 2 position. The enzyme appears to be membrane bound and has a pH optimum around 7, with a requirement for NADPH and oxygen. While trans-cinnamic acid is the apparent substrate for the oxidation, the cis-isomer may be active in some cases (Kindl, 1971; Stoker and Bellis, 1969). In *Melilotus*, the cis-o-coumaric acid is not glucosylated to produce the

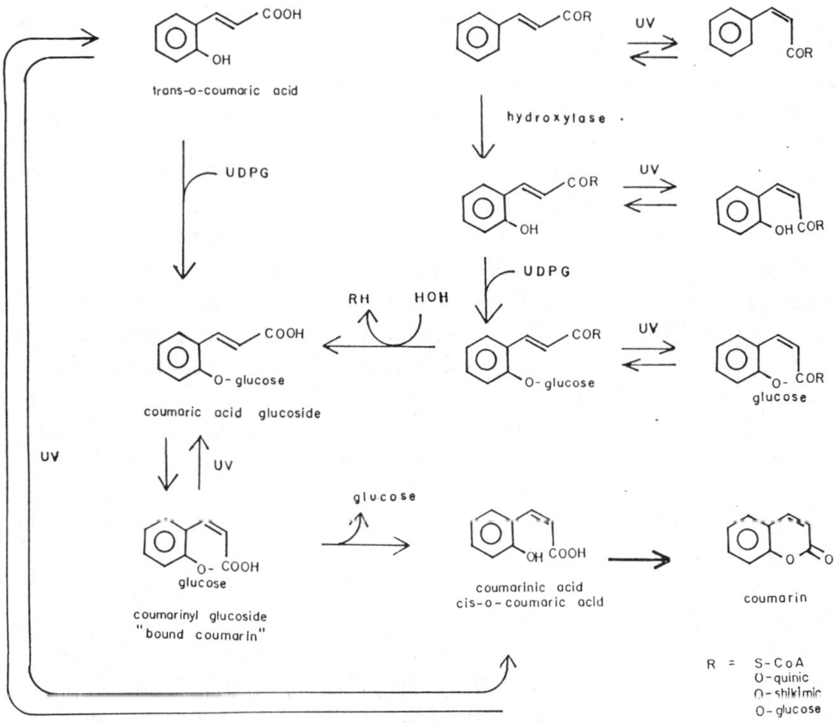

Figure 5. Biosynthetic pathway for the formation of coumarin.

coumarinyl glucosides (Kleinhofs et al., 1964). The cis-isomer can be formed nonenzymatically by UV induced cis-trans isomerization.

Ranjeva et al. (1977) have reported that CA2H as well as PAL are present in thc chloroplasts of *Petunia*, although there was no CA4H in the plastids. On the other hand, no CA2H is found in the microsomes of *Petunia* where the CA4H is localized.

It has been proposed that coumarin is formed in a number of plants only through the loss of glucose from the "bound coumarin," coumarinyl glucoside, followed by spontaneous and possibly nonenzymatic cyclization to coumarin. Under these conditions, free coumarin would arise only when the tissue is disrupted. On the other hand, other plants, e.g., *Dipterx odorata*, contain crystalline coumarin (Murray et al., 1982). If free trans-cinnamic acid is 2-hydroxylated, it should undergo isomerization to form cis-isomers that would cyclize to produce coumarin, but in *Melilotus*, no free coumarin is ever found if care is taken when tissue is extracted (Haskins and Gorz, 1961).

This interesting fact can be explained in at least three different ways. First, any free coumarin formed under normal physiological conditions is quickly and totally metabolized to some other product(s). Second, there is a direct channeling, so that the concentration of free o-coumaric acid is essentially nil. Third, an alternative pathway exists. An example of this is the conjugation of cinnamic acid to some other compound, e.g., glucose, quinic acid, coenzyme A, etc., before 2-hydroxylation. The resultant o-coumaric acid conjugate can then be glucosylated before loss of the protecting group to produce bound coumarin.

The formation of oxidized forms of coumarin apparently occurs through the oxidation of the appropriate hydroxy cinnamic acid at the 2 position rather than the oxidation of the coumarin nucleus. Kindl (1971), for instance, has reported that in *Hydrangea macrophylla*, a plant producing 7-oxidized coumarins, p-coumaric acid was rapidly 2-hydroxylated and lactonized to umbelliferone, while cinnamic acid itself was a poor substrate at best. The enzyme isolated by Kindl differed from that studied by Gestetner and Conn in that it appeared to require a tetrahydropteridine cofactor.

With all of this in mind and with the conflicting results concerning the oxidation of cinnamic acid to either o- or p-coumaric acid, we decided to examine the 2-hydroxylation of cinnamic acid as described by Gestetner and Conn (1974) and confirmed by Brown et al. (1960) and Ranjeva et al. (1977).

Gestetner and Conn based their identification of o-coumaric acid on chromatographic separation of the silylated derivatives and upon recrystallization of radioactively labeled material eluted from thin layer chromatograms. The radioactive products of the enzyme reaction mixture corresponded to two regions of the chromatogram: unreacted cinnamic acid and an area that contained unresolved o- and p-coumaric acid. The coumaric acid region was eluted; authentic o-coumaric acid was added to the eluent, and the mixture was recrystallized. Three recrystallizations gave the same specific radioactivity.

When the intermediates or end products of biosynthetic pathways are relatively stable, recrystallization to constant specific activity is often a method of choice for the determination of components. The labeled compounds are recovered, redissolved and repeatedly recrystallized with authentic unlabeled compounds. If the specific radioactivity of the material remains constant after repeated recrystallizations, it is assumed that the recrystallized material is pure, and that the labeled compound is identical to the compound being recrystallized. However, if the specific activity of the crystallization product decreases after each isolation, then clearly the radioactive compound isolated was not the same as the recrystallized product. Often, the first recrystallization yields a strongly radioactive product, even if the radiolabeled material is different from the compound added to the preparation due to trapping effects of crystals, surface effects, or loose binding of electrostatic or Van der Waals type. Subsequent recrystallizations will reduce the specific radioactivity of the crystalline material until the specific activity approaches zero.

Although recrystallization to constant specific activity is a standard and accepted technique of determining or confirming chemical identity, we decided to check its applicability to this situation.

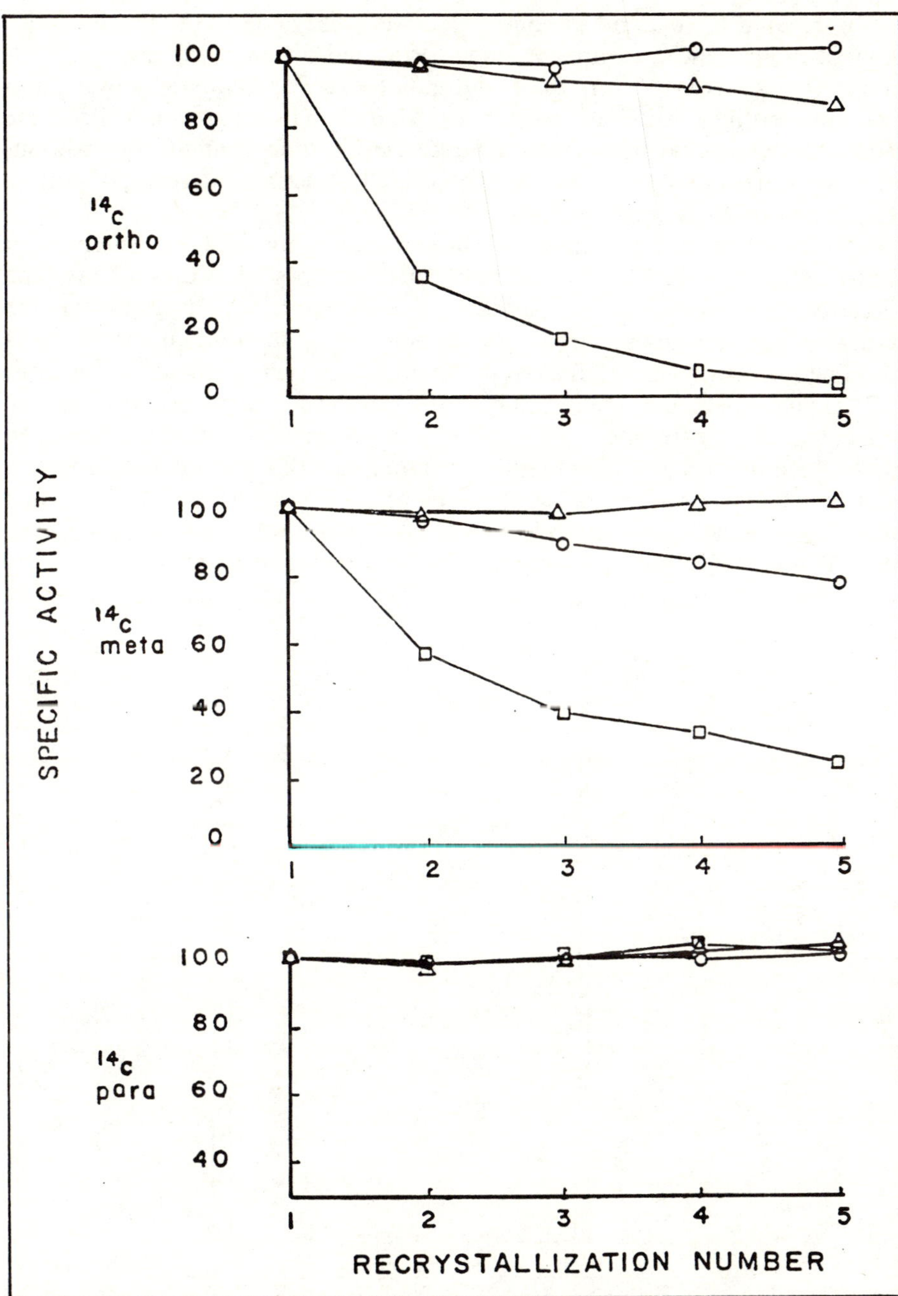

Figure 6. Loss of ^{14}C labeled material from recrystallized coumaric acids. A: loss of ^{14}C-o-coumaric acid added to unlabeled o-coumaric acid, ○-○, unlabeled m-coumaric acid, Δ-Δ, and unlabeled p-coumaric acid, □-□. B: loss of ^{14}C-m-coumaric acid added to unlabeled o-coumaric acid, ○-○, unlabeled m-coumaric acid, Δ-Δ, and unlabeled p-coumaric acid, □-□. C: loss of ^{14}C-p-coumaric acid, ○-○, unlabeled m-coumaric acid, Δ-Δ, or unlabeled p-coumaric acid □-□.

^{14}C-labeled p-, o-, and m-coumaric acids were added to unlabeled solutions of p-, o-, and m-coumaric acids for a total of nine different combinations. The compounds were dissolved in hot water and allowed to recrystallize with cooling. Small aliquots of the dried crystals were weighed and the radioactivity was determined in a liquid scintillation counter. In addition, aliquots were chromatographed by an HPLC system interfaced with a Radiomatic radioisotope detector.

HPLC analyses of the material recovered from the first recrystallization showed the presence of only one radioactive compound in each of the nine mixtures, that of the added radioactive compound, which confirmed the purity of the starting material. As shown in Figure 6 all of the samples contained considerable radioactivity in the first recrystallization fraction. HPLC analysis clearly identified both the radioactive coumaric acid and the unlabeled UV absorbing cold carrier. For example, Figure 7 shows an HPLC tracing of the first and last recrystallization of radioactive o-coumaric acid that had been recrystallized in unlabeled o-coumaric acid. Note that both the radioisotope peak and the UV absorbing peak overlap at 15 minutes. When ^{14}C labeled o-coumaric acid was added to unlabeled o, m, and p-coumaric acids, the specific activity decreased with meta and para but

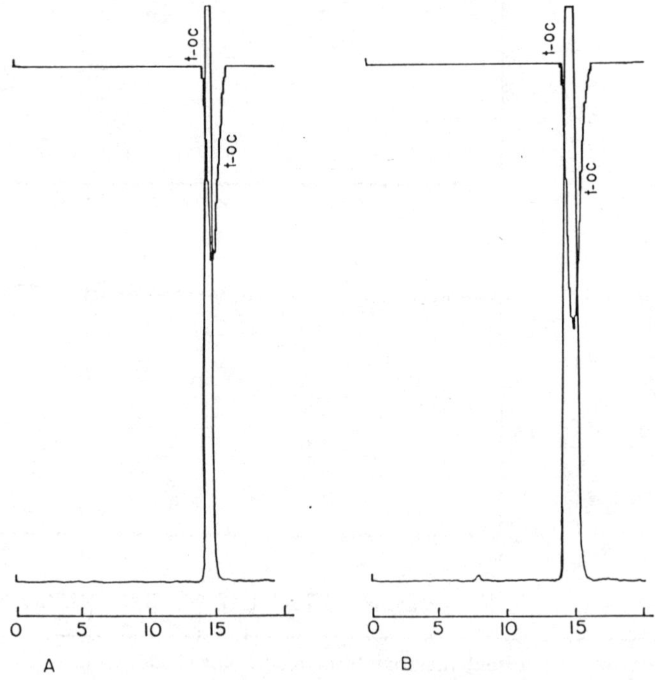

Figure 7. HPLC analysis of recrystallized ^{14}C o-coumaric acid in unlabeled o-coumaric acid. UV detector on bottom tracing and radioisotope detector inverted signal on top. A: ^{14}C o-coumaric acid added to unlabeled o-coumaric acid, first recrystallization, B: fifth recrystallization, shows no change in ratio of UV absorbance to radiosotope detection.

remained constant with ortho-coumaric acid. Similar results were obtained when ^{14}C labeled m-coumaric acid was added to each of the three cold carriers; the m-coumaric acid retained a constant specific activity while the radioactivity in the o- and p-coumaric acids decreased. Figure 8 shows clearly that the radioisotopic peak decreased significantly even when the injection volume of the cold p-coumaric acid was increased 20-fold. However, when the ^{14}C-labeled p-coumaric acid was added to each cold carrier, all fractions retained a high specific activity through all recrystallizations. This result was unexpected, as recrystallization to constant specific activity is a standard procedure for chemical identification. HPLC analysis of first and last recrystallizations of radioisotopic p-coumaric acid in o-coumaric acid confirmed the identity of the recrystallized products, i.e., the UV absorbing peak is due to the o-coumaric acid but the radioactivity is due to p-coumaric acid (Fig. 9). We repeated these experiments using both aqueous acetone and aqueous methanol as recrystallizing solvents with essentially the same results (data not shown). Care was taken to avoid precipitation of the coumaric acids during the recrystallizations, and by the fifth recrystallization approximately 90% of the weight of the starting material

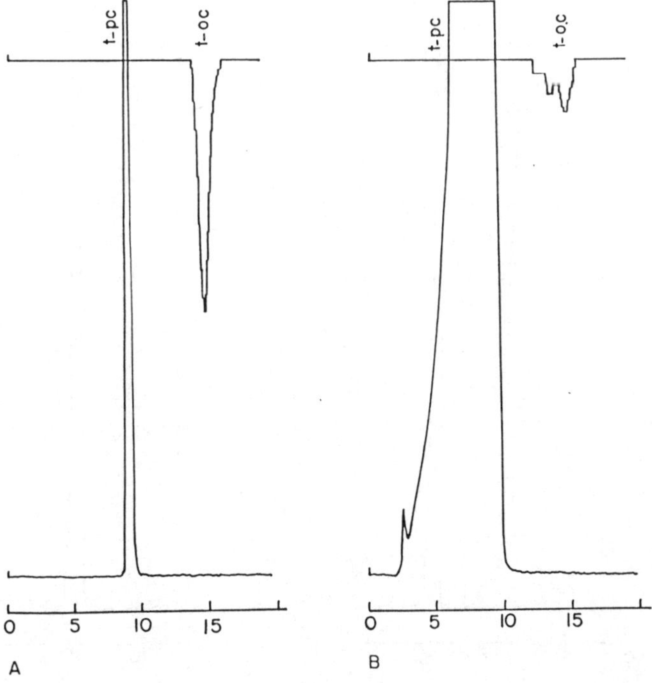

Figure 8. HPLC analysis of recrystallized ^{14}C o-coumaric acid in unlabeled p-coumaric acid. UV detector on bottom tracing and radioisotope detector inverted signal on top. A: ^{14}C o-coumaric acid added to unlabeled p-coumaric acid, first recrystallization. B: fifth recrystallization, radioactivity has decreased with respect to UV absorbance. UV absorbance peak and the radioactivity peak do not coelute.

had been lost. All colored contaminants present in the original unlabeled samples were eliminated by the fifth recrystallization. When the recrystallized samples were analyzed by HPLC, the radioactive compounds were clearly separated from the cold carrier and eluted with predictable retention times.

The solubilities of the coumaric acids are similar and relatively high in hot water. The p-isomer is, if anything, slightly more soluble in aqueous methanol than the o-isomer. This should reduce even further the possibility that the radiolabeled p-isomer was rapidly and completely precipitated in the crystallizing o-coumaric acid. To maximize this advantage we used large quantities of solvent and recrystallized small quantities of the o-coumaric acid. In these circumstances, only 60% of the total radioactivity was recovered in each step, while the specific activity remained constant.

The implication of this study is that the technique of recrystallization to constant specific radioactivity may not be accepted as a standard technique for the identification of metabolites from complex mixtures. Even if preliminary experiments are used to determine whether the radiolabeled precursor can be trapped by the proposed end product, there can be no assurance that the labeled compound will not be metabolically converted to an inter-

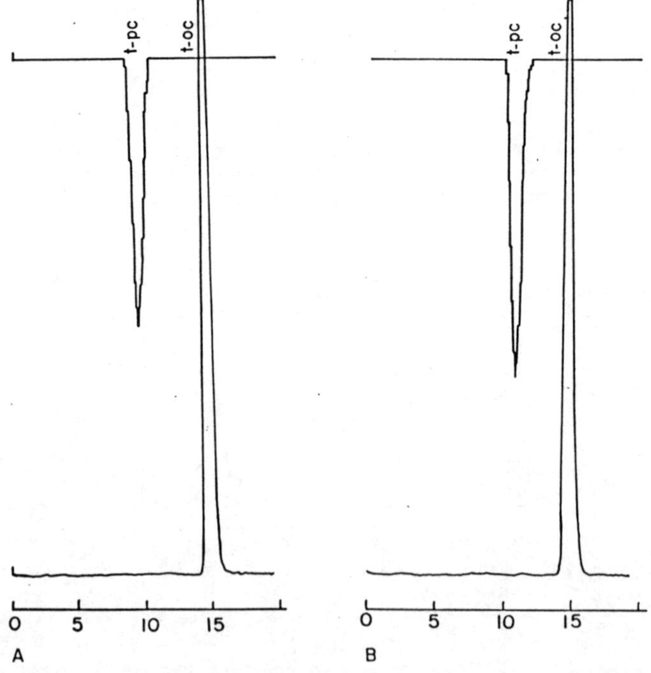

Figure 9. HPLC analysis of recrystallized ^{14}C p-coumaric acid in unlabeled o-coumaric acid. A: first recrystallization. B: fifth recrystallization. Although the radioactivity does not coelute with the UV absorbance, the ratio of radioactivity to UV absorbance remains constant.

mediate compound, which may precipitate out with a presumed end product, and give a constant specific activity through repeated recrystallizations.

This work convincingly shows that recrystallization to constant specific activity, even when done in a variety of solvents, as in this case, gave erroneous results concerning the identification of these compounds. Although this procedure was limited to three common isomers of coumaric acid, it raises doubts concerning the reliability of recrystallization as an analytical procedure. In this case, the use of the radioisotope detector gave unquestioned identification of ortho or para-hydroxylation of cinnamic acid to resolve conflicting reports in the literature. Researchers are cautioned, so that their results are not compromised by similar, biochemically precise, but faulty recrystallization data.

CIS-TRANS ISOMERIZATION OF CINNAMIC AND COUMARIC ACIDS

As indicated previously, glycosides of coumaric acids undergo a cis-trans isomerization in the presence of UV light. This isomerization can lead to the misidentification of the phenylpropanoid intermediates because of changes in retention times on chromatographic systems. To address this problem we exposed ^{14}C radiolabeled cinnamic and coumaric acids to UV light.

An authentic sample of ^{14}C trans-cinnamic acids was prepared from ^{14}C-phenylalanine, purified by ion chromatography, and irradiated with UV light. As expected, a cis-trans-isomerization occurred. The cis-cinnamic acid eluted very early in our HPLC system with a retention time of 13 minutes (Fig. 10). We also irradiated the o-, m-, and p-coumaric acid with UV light and showed that each of the cis-trans-isomers could be separated with the HPLC system (Fig. 11 & 12).

The HPLC chromatographs show that it is possible to misidentify cis-cinnamic acid as either trans-o-coumaric acid or coumarin. Since this misidentification is due to coelution of unexpected isomers that occurred within our chromatographic system, it is possible that similar misidentifications may have occurred with previous workers. UV irradiation, intentional or accidental can lead to multiple substrates and products, which may not be stable compounds.

For example, when trans-o-coumaric acid was irradiated with UV light, three peaks occurred on the HPLC chromatogram. One corresponded to the original trans-isomer, the second to the cis-isomer, and a third smaller peak to coumarin. As this solution aged in the dark, the concentration of coumarin increased with a concommitant decrease in the concentration of cis-o-coumaric acid (Fig. 12). The level of trans-o-coumaric remained constant in the dark. It is apparent that these unstable intermediates could effect the analysis of various components and lead to variable results. In general, the literature, although cognizant of the possibility of cis-trans-isomerization, tends to ignore the difficulties that may arise in chromatography from these isomer pairs. Retention times of standards often appear to consider only the trans-isomers.

SEPARATION OF STEREO-ISOMERS ON HPLC

The use of radioisotope detectors usually results in a decrease in the resolution of the HPLC system. To insure the authenticity of the cis-cinnamic acid, radioisotopically labeled ^{14}C phenylalanine, together with labeled cis- and trans-cinnamic acid were separated on the HPLC system (Fig. 10A). The cis-isomer of cinnamic acid was collected and exposed to UV irradiation, causing a cis-trans isomerization of irradiated material and, once again, gave two peaks coeluting exactly with cis- and trans-cinnamic acid (Fig. 10B). The use of the cis-trans isomerization coupled with HPLC and radioisotope detection, therefore, provided a mechanism for confirming identification of cinnamic acid even when the coumaric acid elution profile overlapped cinnamic acid due to a loss of resolution.

EXPERIMENTAL PROCEDURES

^{14}C-U-Phenylalanine and ^{14}C-U-tyrosine were purchased from New England Nuclear with specific activity of 250 mCi/mMole; ^{14}C-2-Malonic acid was also purchased from New England Nuclear with specific activity of 500 mCi/mMole.

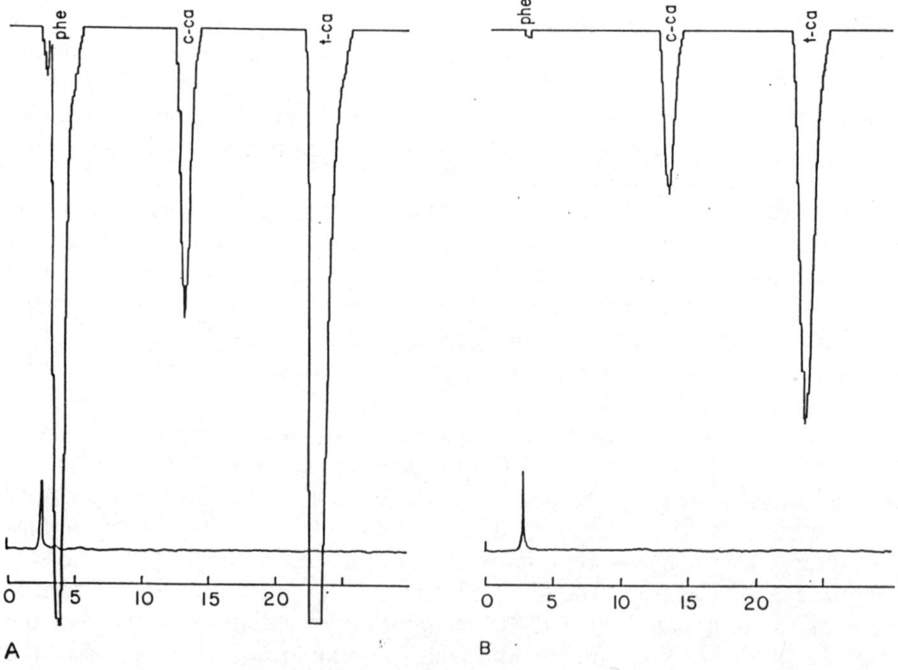

Figure 10. HPLC analysis of ^{14}C phenylalanine, ^{14}C-cis-cinnamic acid and ^{14}C-trans-cinnamic acid. A: Although the radioisotope detector shows full scale deflection of the radioactive peaks, no UV absorbance at 254 nm is detected. B: cis-cinnamic acid was collected, exposed to UV light for one hour and rechromatographed to show cis-trans-isomerization.

HPLC APPARATUS

The HPLC system was composed of a 10 μ reverse phase C18 column with an isocratic mobile phase of 40% methanol containing 0.1% acetic acid. A Waters Model 440 UV detector was used at 254 nm to monitor the elution of the test compounds at a solvent flow of 1.2 ml/min. To monitor low level radioactive intermediates, an on-line radioisotope detector Radiomatic FLO-ONE Model HS was used with a 2.5 ml flow-through scintillant chamber. The approximate efficiency of the radioisotope detector using FLO-SCINT III scintillation cocktail with ^{14}C isotopes was measured 61% with a scintillant/solvent ratio of 3:1.

PAL ASSAY

Phenylalanine ammonia lyase activity was examined in four different plant sources, *Heirochloe odorata*, *Melilotus alba*, *Hydrangea*, and *Petunia*.

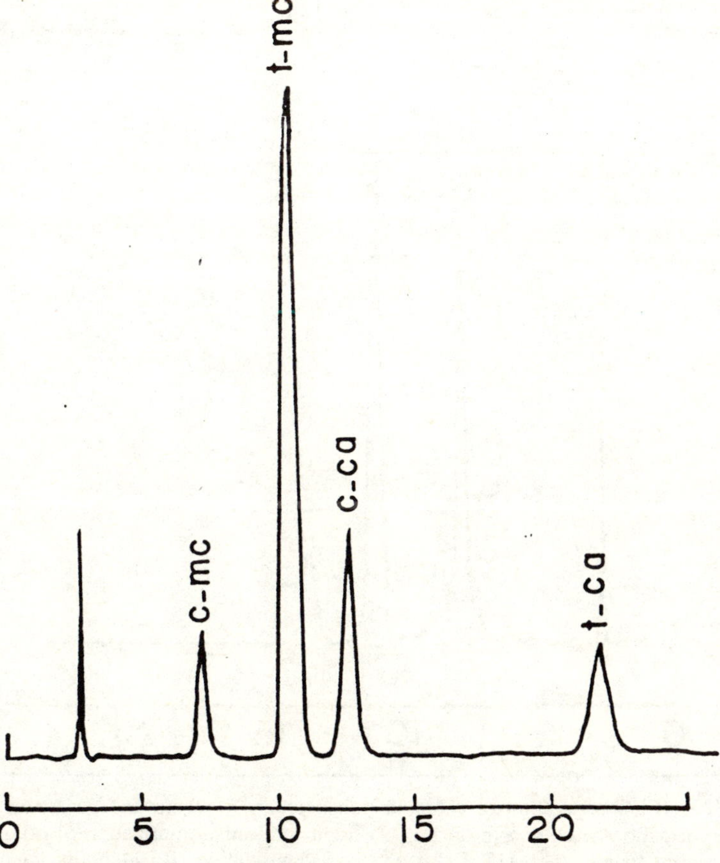

Figure 11. HPLC separation of cis- and trans-isomers of meta-coumaric acid and cinnamic acid using UV detector at 254 nm.

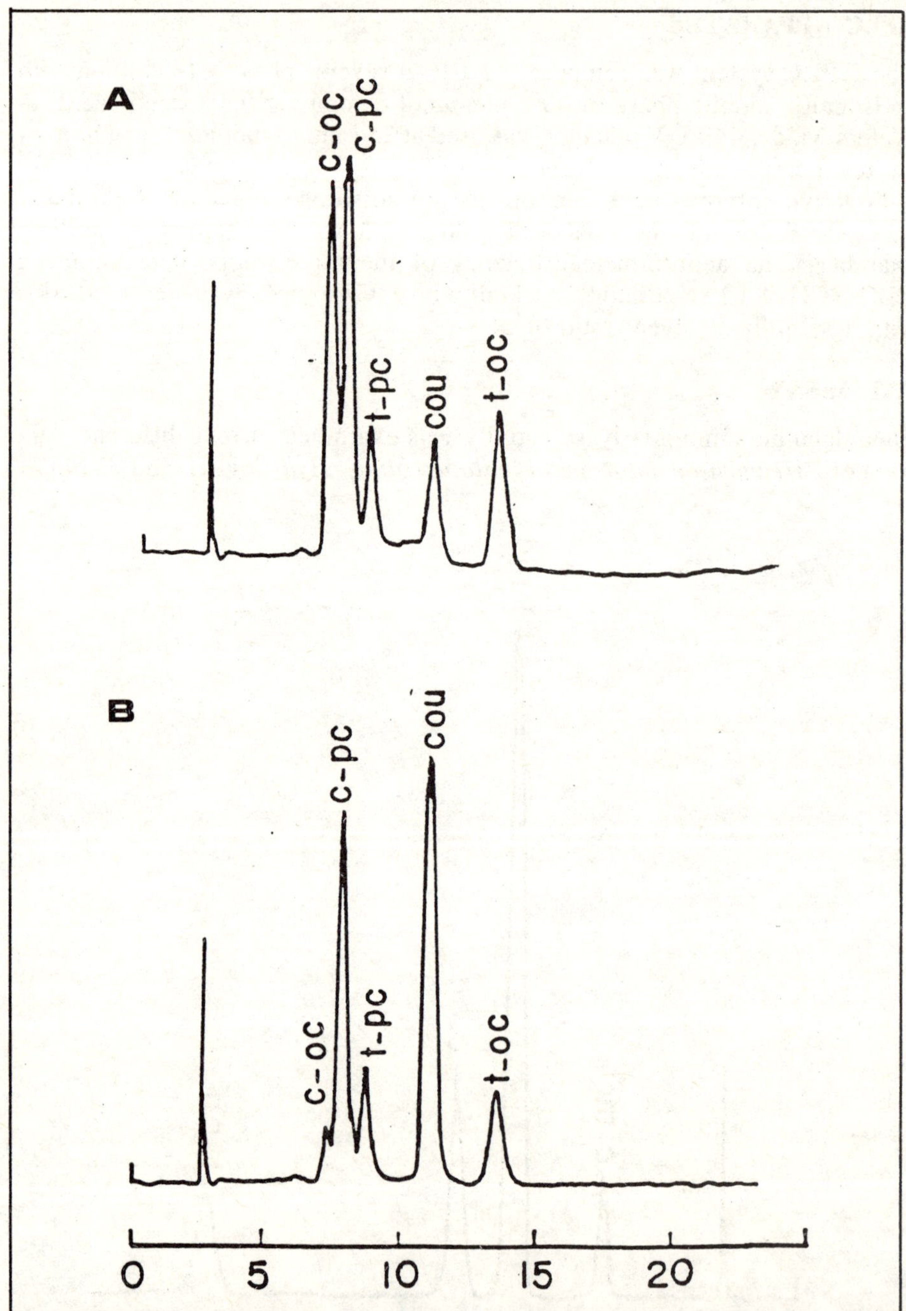

Figure 12. HPLC separation of cis- and trans-isomers of ortho and para-coumaric acid. Note the formation of coumarin during UV light induced isomerization of ortho-coumaric acid. A: Immediately after UV light induced isomerization. B: 24 hours after UV light induced isomerization. Note the increase in coumarin and decrease in cis-ortho-coumaric acid.

Approximately 3 grams of plant material was ground in 10 ml of borate buffer and the pH of the filtered homogenate was adjusted to 8.8. ^{14}C-U-Phenylalanine was added to the homogenate and the mixture was incubated at 37°C for two hours. The enzymatic reaction was stopped with the addition of acidic ethanol. After centrifugation and 0.45 micron ultrafiltration a 15 μl aliquot was injected onto a C18 reverse phase HPLC column, using 40% MeOH/0.1% acetic acid, 1.2 ml/min as mobile phase. The eluant was monitored both with UV detector at 254 nm and with a radioisotope detector. Retention times had been previously determined with unlabeled compounds.

TAL ASSAY

Tyrosine ammonia lyase activity was determined in a manner analogous to that used for PAL analysis with the addition of ^{14}C-U-tyrosine in place of the phenylalanine. The commercially available labeled tyrosine was contaminated with a component accounting for 3% of the total radioactivity.

CARBON-14 LABELED ACID SYNTHESES

Radioactive trans-m-coumaric acid was made coupling ^{14}C-2-malonic acid with m-hydroxybenzaldehyde essentially by the method of Adams and Bockstahler (1952). The glucoside of ^{14}C-2-trans-o-coumaric acid was treated with almond emulsin to release the free acid. This glucoside of ^{14}C-2-trans-o-coumaric acid was prepared by the Knovenagel synthesis with labeled malonic acid and helicin, the glucoside of salicylaldehyde, by the method of Helferich and Lutzman (1938). ^{14}C-U-Trans-p-coumaric acid was prepared both by the deamination of tyrosine with nonspecific PAL and by condensation of labeled malonic acid and p-hydroxybenzaldehyde. ^{14}C-U-Trans-cinnamic acid was prepared by treating ^{14}C-U-phenylalanine with PAL at pH 8.5 for one hour at 30°C. The enzymatic reaction was stopped with the addition of carrier cinnamic acid in acidified ethanol. The solution was Millipore filtered to remove precipitated proteins. All of the ^{14}C labeled acids were in the thermodynamically favorable trans-configuration. The ^{14}C-labeled cinnamic acid was purified first by ion chromatography followed by reverse phase HPLC. The ^{14}C-labeled coumaric acids used in the recrystallization experiments were used without further purification. Cis-trans-isomerization was effected by allowing a solution of the free trans-acids in aqueous methanol to equilibrate under UV lights (254 and 280 nm) for one hour.

CINNAMIC ACID HYDROXYLASE ASSAYS

The CA2H and CA4H activities were monitored by the addition of NADPH and an NADPH regenerating system of glucose-6-phosphate and glucose-6-phosphate dehydrogenase. ^{14}C-U-Cinnamic acid in the trans-configuration as prepared above was used as substrate. A crude plant homogenate was used as the enzyme source. The assay was run at both pH 5.8 and pH 8.4 for one hour at 30°C. Acidified ethanol and unlabeled carrier acids were added to stop incubation. Purification procedures and chromatography were identical to the PAL and TAL assays.

Abbreviations used in Figures:

phe = phenylalanine	PAL = phenylalanine ammonia lyase
tyr = tyrosine	TAL = tyrosine ammonia lyase
t-ca = trans-cinnamic acid	CA2H = cinnamic acid-2-hydroxylase
t-pc = trans-p-coumaric acid	CA4H = cinnamic acid-4-hydroxylase
t-mc = trans-m-coumaric acid	c-mc = cis-m-coumaric acid
cou = coumarin	c-ca = cis-cinnamic acid
t-oc = trans-o-coumaric acid	c-oc = cis-o-coumaric acid
t-cf = trans-caffeic acid	c-pc = cis-p-coumaric acid

REFERENCES

Adams, R. and Bockstahler, T.E. (1952) Preparation and reactions of o-hydroxycinnamic acids and esters. *Ann. Biochem.* 74:5346-5348

Alibert, G., Ranjeva, R. and Boudet, A.M. (1977). Organisation subcellulaire des voise de synthese des composes phenoliques. *Physiol. Veg.* 15:279-301.

Alibert, G., Ranjeva, R. and Boudet, A.M. (1972a). Recherches sur les enzymes catalysant la formation des acides phenoliques chez *Quercus pedunculata* II. Localisation intracellulaire de la phenylalanine ammoniaque-lyase, de la cinnamate 4-hydroxylase, et de la "benzoate synthase." *Biochem. Biophys. Acta.* 279: 282-289.

Alibert, G., Ranjeva, R. and Boudet, A.M. (1972b). Recherches sur les enzymes catalysant la biosyntheses acides phenoliques chez *Quercus pedunclata*. III. Formation sequentielle, a partir de la phenylalanine, des acides cinnamique, p-coumarique et cafeique, par des organites cellulaires isoles. *Physiol. Plant.* 27: 240-243.

Albu, E., Sprichez, C. and Dabala, I. (1969). Action of cinnamate on rise rhizongenesis in the hydrangea (*Hydrangea hortensis*). Stud. Univ. Babes-Bolyai, *Sec. Biol.* 14:83.

Bartl, K., Cavalar, C., Krebs, T., Ripp, E., Retey, J., Hull, W.E., Gunther, H. and Simon, H. (1977). Synthesis of stereospecifically deuterated phenylalanines and determination of their configuration. *Eur. J. Biochem.* 72:247-250.

Beaudoin-Eagan, L.D, and Thorpe, T.A. (1985). Tyrosine and phenylalanine ammonia lyase activities during shoot initiation in tobacco callus cultures. *Plant Physiology* 78:438-441.

Bell, E.A. and Charlwooded, B.V. (1980). Secondary Plant Products. In: *Encyclopedia of Plant Physiology New Series*. Vol. 8, Springer-Verlag, New York.

Bell, E.A. (1981). The physiological role(s) of secondary (natural) products. In: *The Biochemistry of Plants* Vol. 7 Secondary Plant Products, E.E. Conn (ed). Academic Press. New York. pp. 1-20.

Benveniste, I., Salaun, J.P. and Durst, F. (1978). Wounding-induced cinnamic acid hydroxylase in Jerusalem artichoke tuber. *Phytochemistry* 16:69-73.

Billet, E.E. and Smith, H. (1980). Control of phenylalanine ammonia-lyase and cinnamic acid 4-hydroxylase in gherkin tissue. *Phytochemistry* 19:1035-1041.

Birch, A.J., Massy-Westropp, R.A. and Moye, C.J. (1955). Studies in relation to biosynthesis. VII. 2-hydroxy-6-methylbenzoic acid in penicillium. *Aust. J. Chem.* 8:539-544.

Birch, A.J., Ryan, A.J. and Smith, H. (1958a). Studies in relation to biosynthesis. XIX. The biosynthesis of helminthosporin. *J. Chem. Soc.* 4773-4774.

Birch, A.J., Massy-Westropp, R.A., Rickards, R.W. and Smith, H. (1958b). Studies in relation to biosynthesis. XV. Origin of terpenoid structures in mycelianamide and mycophenolic acid. *J. Chem. Soc.* pp. 369-375.

Brown, S.A., Towers, G.H.N. and Wright, D. (1960). Biosynthesis of the coumarins. Tracer studies on coumarin formation in *Hierochloe odorata* and *Melilotus officinalis*. *Can J. Biochem. Physiol.* 38:143-156.

Buche, T. and Sandermann, H. (1973). Lipid dependence of plant microsomal cinnamic acid 4-hydroxylase. *Arch. Biochem. and Biophys.* 158: 445-447.

Camm, E.L. and Towers, G.H.N. (1973a). Is cinnamic acid 4-hydroxylase of potato a light-induced enzyme. *Can. J. Bot.* 51: 824-825.

Camm, E.L. and Towers, G.H.N. (1973b). Phenylalanine ammonia-lyase. *Phytochemistry* 12: 961-973.

Camm, E.L. and Towers, G.H.N. (1977). Phenylalanine ammonia-lyase. In: *Progress in Phytochemistry*. Vol. 4, L. Reinhold, J.B. Harborne & T. Swain, (eds). Pergamon Press, New York. 4: 169-188.

Collie, J.N. (1893). The production of naphthalene derivatives from dehydracetic acid. *J. Chem. Soc.* 63:329-337.

Collie, J.N. (1907). Derivatives of the multiple keten group. *J. Chem. Soc.* 91:1806-1831.

Court, W.A. (1977). *J. Chromatogr.* 166:271-279.

Czichi, U. and Kindl, H. (1977). Formation of hydroxycinnamic acids from L-phenylalanine on thylakoids of *Dunaliella marina*. Hoppe-Seyler's Z. *Physiol. Chem.* 356: 475-485.

Davis, B.D. (1958). On the importance of being ionized. *Arch. Biochem. Biophys.* 78:497-509.

Ellis, B.E. and Amrhein, N. (1971). The 'NIH-shift' during aromatic ortho-hydroxylation in higher plants. *Phytochemistry* 10:3069-3072.

Gatenbeck, S. and Mosbach, K. (1959). Acetate carboxyl oxygen (^{18}O) as donor for phenolic hydroxy groups of orsellinic acid produced by fungi. *Acta. Chem. Scand.* 13:1561-1564.

Gestetner, B. and Conn, E.E. (1974). The 2-hydroxylation of trans-cinnamic acid by chloroplasts from *Melilotus alba Desr. Arch. Biochem. Biophys.* 163: 617-624.

Givot, I.L., Smith, T.A. and Abeles, R.H. (1969). Studies on the mechanism of action and the structure of the electrophilic center of histidine ammonia lyase. *Biol. Chem.* 244: 6341-6353.

Gross, G.G., Stockigt, J., Mansell, R.L. and Zenk, M.H. (1973). Three novel enzymes involved in the reduction of ferulic acid to coniferyl alcohol in higher plants: Ferulate; CoA ligase, feruloyl CoA reductase and coniferyl alcohol oxidoreductase. *FEBS Lett.* 31:283-286.

Guerra, D.A., Anderson, A.J. and Salisbury, F.B. (1985). Reduced phenylalanine ammonia-lyase and tyrosine ammonia-lyase activities and lignin synthesis in wheat grown under low pressure sodium lamps. *Plant Physiology* 78: 126-130.

Hanson, K.R. and Havir, E.A. (1969). Reduction of the active site of L-phenylalanine ammonia-lyase. *Fed. Proc., Fed. Am. Soc. Exp. Biol.* 28:602.

Hanson, K.R. and Havir, E.A. (1972). *The Enzymes*. P.D. Boyer (ed). Academic Press Inc. New York. 7:75-166.

Hanson, K.R. and Havir, E.A. (1981). *The Biochemistry of Plants*. Vol. 7, E.E. Conn (ed). Academic Press Inc. New York. pp. 577-625.

Harborne, J.B. (1967). *Comparative Biochemistry of the Flavonoids*. Academic Press, New York.

Harborne, J.B. (1980). Plant phenolics. Secondary plant products. Vol. 8, In: *Encyclopedia of Plant Physiology*, Springer-Verlag. New York. pp. 329-402.

Harborne, J.B. and Mabry, T.J. (1982). *The Flavonoids, Advances in research*. Chapman and Hall, London.

Haskin, F.A. and Gorz, H.J. (1961). A reappraisal of the relationship between free and bound coumarin in *Melilotus*. *Crop Sci.* 1:320-322.

Havir, E.A. (1979). Binding of polysaccharide by L-phenylalanine ammonia-lyase (Maize). *Fed. Proc., Fed. Am. Soc. Exp. Biol.* 38:354.

Havir, E.A. and Hanson, K.R. (1973). L-phenylalanine ammonia-lyase (Maize and Potato). Evidence that the enzyme is composed of four subunits. *Biochemistry* 12:1583-1591.

Havir, E.A., Reid, P.D. and Marsh, H.V.J. (1971). L-phenylalanine ammonia-lyase (Maize) evidence for a common catalytic site for L-phenylalanine and L-tyrosine. *Plant Physiol.* 48:130-136.

Helferich, B. and Lutzman, H. (1938). Glucoside von Phenol-carbonsaure, ihre fermentative Spaltung and ihre Selbstzersetzung. *ANN.* 537:11-21.

Higuchi, T., Shimada, M., Nakatsubo, F. and Tanahashi, M. (1977). Differences in biosynthesis of guaracyl and syringyl lignans in wood. *Wood Sci. Technol.* 11:153-167.

Hill, A.C. and Rhodes, M.J.C. (1975). The properties of cinnamic acid 4-hydroxylase of aged Swede root disks. *Phytochemistry* 14:2387-2391.

Hrazdina, G., Wagner, G.J. and Siegleman, H.W. (1978). Subcellular localization of enzymes of anthocyanin biosynthesis in protoplasts.*Phytochem.* 17:53-56.

Kindl, H. (1971). Zur frage der ortho-Hydroxylierung aromatischer carbonsauren-in hoheren pflanzen. Hoppe-Seyler's Z. *Physiol. Chem.* 353:78-84.

Kleinhofs, A., Haskins, F.R. and Giorz, H.J. (1964). Trans-o-hydroxycinnamic acid glucosylation in cell-free extracts of *Melilotus alba. Phytochem.* 6:1313- 1318.

Kojima, M and Uritani, T. (1972). Studies on the mechanism of action biosynthesis in sweet potato root tissue using trans-cinnamic acid-2-^{14}C and quinic acid - G - ^{3}H. *Plant and Cell Physiol.* 13:311-319.

Koukol, J. and Conn, E.C. (1961). The metabolism of aromatic compounds in higher plants. *J. Biol. Chem.* 236:2692-2698.

Kreuzler, F. and Hahlbrock, K. (1972). Enzymatic synthesis of aromatic compouds in the higher plants: Formation of naringenin (5,7,4'-trihydroxy flavanone) from p-coumaroyl coenzyme A and malonyl coenzyme A. *FEBS Lett.* 288:69-72.

Kreuzler, F. and Hahlbrook, K. (1975). Enzymatic synthesis of aromatic compounds in higher plants. Formation of bis-noryangonin (4-hydroxy-6(4-hydroxystyryl)-2-pyrone) from p-coumaroyl coenzyme A and malonyl coenzyme A. *Arch. Biochem. Biophys.* 169:84-90.

Krstulovic, A.M. and Brown, P.R. (1982). *Reversed-phase high-performance liquid chromatography*. Wiley Press, New York.

Levy, C.C. and Zucker, M. (1960). Cinnamyl and p-coumaryl esters as intermediates in the biosynthesis of chlorogenic acid. *J. Biol. Chem.* 235:2418-2425.

Mathews, J.M. and Ortiz de Montellano, P.R. (1982). Autocatalytic inactivation of plant cytochrome P-450 enzymes: Selective inactivation of cinnamic acid 4-hydroxylase for *Helianthus tuberosus* by l-aminobenzotriazole. *Arch. Biochem. and Biophys.* 216:522-529.

McClure, J. (1979). The physiology of phenolic compounds in plants. *Recent Adv. Phytochemistry.* 12:525-556.

Moskowitz, A.H. and Hrazdina, G. (1981). Vacuole contents of fruit subepidermal cells from *Vitis* species. *Plant Phys.* 68:686-692.

Murray, R.D.H., Mendet, J. and Brown, S.A. (1982). *The Natural Coumarins*. John Wiley and Sons, New York.

Nair, P.M. and Vining, L.C. (1965). Cinnamic acid hydroxylase in spinach. *Phytochemistry* 4:161-168.

Neish, A.C. (1961). Formation of m- and p-coumaric acids by enzymatic deamination of the corresponding isomers of tyrosine. *Phytochemistry* 1:1-24.

Nishizawa, A.N., Wolosiuk, R.A. and Buchanan, B.B. (1979). Chloroplast phenylalanine ammonia-lyase from spinach leaves. *Planta* 145:7-12.

Poulton, J.E., McRee, D. Conn, E.E., Saunders, J.A., Blume, D.E. and McClure, J.W. (1980). Reappraisal of the inhibitory effect of certain sugars used as osmotica on phenylalanine ammonia-lyase activity. *Phytochem.* 19:775-777.

Ranjeva, R., Alibert, G. and Boudet, A.M. (1977). Metabolisme des composes phenoliques chez le *Petunia V.* Utilisation de la phenylalanine par des chloroplastes isoles. *Plant Sci.* 10:225-234.

Ranjeva, R., Boudet, A.M. and Alibert, G. (1979). *FEBS-Symp.* 55:91.

Rhodes, M.J.C. and Wooltorton, L.S.C. (1976). The enzymatic conversion of hydroxycinnamic acids to p-coumaroyl, quinic, and chlorogenic acid in tomato fruits. *Phytochemistry* 15:947-951.

Rich, P.R. and Lamb, C.J. (1977). Biophysical and enzymological studies upon the interaction of trans-cinnamic acid with higher plant microsomal phytochrome P-450. *Eur. J. Biochem.* 72:353-360.

Russell, D.W. (1971). The metabolism of aromatic compounds in higher plants. *J. Biol. Chem.* 246:3870-3878.

Russell, D.W. and Conn, E.E. (1967). The cinnamic acid 4-hydroxylase of pea seedlings. *Arch. Biochem. Biophy.* 122:256-258.

Sato, M. (1966). Metabolism of phenolic substances by the chloroplasts - II. Conversion by the isolated chloroplasts of p-coumaric acid to caffeic acid. *Phytochemistry* 5:385-389.

Saunders, J.A., Conn, E.E., Lin, C.H. and Shimada, M. (1977). Localization of cinnamic acid 4-mono-oxygenase and the membrane-bound enzyme system for dhurrin biosynthesis in *Sorghum* seedling. *Plant Physiol.* 60:629-634.

Saunders, J.A. and McClure, J.W. (1974). The suitability of a quantitative spectrophotometric assay for phenylalanine ammonia-lyase activity in barley, pea, and buckwheat seedlings. *Plant Phys.* 54:412-413.

Saunders, J.A. and McClure, J.W. (1975). Phytochrome controlled phenylalanine ammonia lyase in *Hordeum vulgare* plastids. *Phytochemistry* 14:1285-1289.

Saunders, J.A. and McClure, J.W. (1976a). The occurrence and photoregulation of flavonoids in barley plastids. *Phytochemistry* 15:805-807.

Saunders, J.A. and McClure, J.W. (1976b). The distribution of flavonoids in chloroplasts of twenty-five species of vascular plants. *Phytochemistry* 15:809-810.

Sisson, V.A. and Saunders, J.A. (1982). Alkaloid composition of the USDA Tobacco (*Nicotiana tabacum L.*). Introduction Collection. *Tobacco Science* 24.117-120.

Stafford, H.A. (1969). Changes in phenolic compounds and related enzymes in young plants of *Sorghum*. *Phytochem.* 8:743-752.

Stafford, H.A. (1974). Possible multi-enzyme complexes regulating the formation of C_6-C_3 phenolic compounds and lignins in higher plants. *Recent Adv. Phytochem.* 8:53-79.

Stallknecht, G.F. (1972). Coumarin-induced tuber formation on excised shoots of *Solanum tuberosum L.* cultured *in vitro*. *Plant Physiol.* 50:412-413.

Stoker, J.R. and Bellis, D.M. (1962). The biosynthesis of coumarin in *Melilotus alba*. *J. Biol. Chem.* 237:2303-2305.

Swain, T., Harborne, J.B. and VanSumere, C.F. (1979). Biochemistry of plant phenolics. *Rec. Adv. Phytochemistry* 12:1-641.

Ultizur, S. and Poljakoff-Mayber, A. (1963). Oxidative phosphorylation in germinating lettuce seeds. *J. Exp. Bot.* 14:95-100.

Zahringer, U., Ebel, J. and Grisebach, H. (1978). Induction of phytoalexin synthesis in soybean elicitor-induced increase in enzyme activities of flavonoid biosynthesis and incorporation of mevalonate into Glyceollin. *Arch. Biochem. Biophys.* 188:450-455.

Zenk, M.H. (1978). Recent work on cinnamoyl coenzyme A derivatives. *Rec. Adv. Phytochem.* 12:139-176.

Progress in HPLC, Vol. 3, pp. 191-210
Parvez et al. (Eds)

HPLC equipped with synchronized accumulating radioisotope detector and its application to the study of drug metabolism

SHIGEO BABA
Tokyo College of Pharmacy, Hachioji, Tokyo, Japan

INTRODUCTION

In radioisotope tracer experiments, thin-layer chromatography (TLC) and paper chromatography have generally been used for the separation of radioactive compounds. However, there have been few applications of gas chromatography (GLC) and high performance liquid chromatography (HPLC) for such experiments, although these two techniques often exhibit a higher chromatographic resolution. An inherent drawback associated with a conventional radioisotope detector using a single counting cell is that the detecting efficiency can not be improved without sacrificing the resolution. To overcome this drawback, we have devised a synchronized accumulating radioisotope detector (SARD), which was successfully used for TLC (Baba *et al.*, 1979) and GLC (Baba and Kasuya, 1980). This chapter describes the application of this detector for HPLC (Baba *et al.*, 1982, 1987).

WORKING PRINCIPLE OF A SYNCHRONIZED ACCUMULATING RADIOISOTOPE DETECTOR

The volume of detectors used for GLC (FID, ECD, etc.), or HPLC (UV detector) is negligibly small compared with the volume emerging from a chromatographic column. On the other hand, for radioactivity measurement, a detector having much larger volume must be used to obtain high sensitivity. For detecting radioactivity in an eluate from a chromatographic column, the conventional detector assembly is composed of a single detector cell. However, this type of detector does not permit a simultaneous improvement in both the detecting efficiency for the radioactivity and the resolution. This is schematically illustrated in Figure 1. In this figure, two counting cells, through which two radioactive components (X and Y) travel at the flow rate (ml/min) of u, are considered. The volumes (ml) of the counting cells I and II are v and $3v$, respectively. Suppose a bulk of radioactive components occupies almost the same volume as the cell I volume. Retention time unit, a (sec), is given by $60v/u$. Although a counting cell

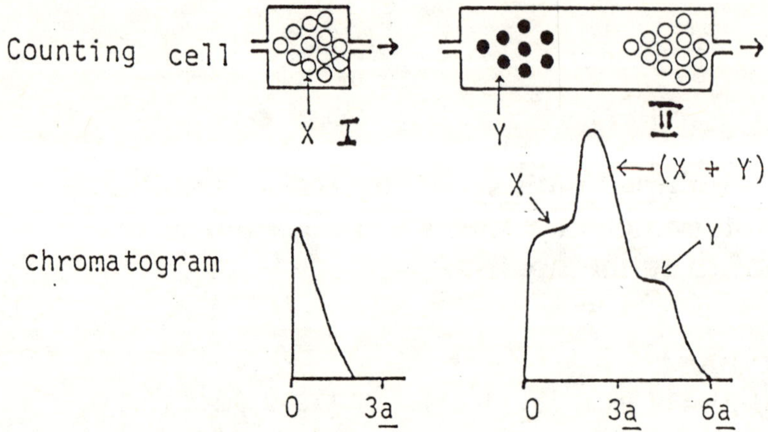

Figure 1. Inherent drawback of the conventional method

having smaller volume provides a sharper radioactivity peak, the transit time of the component X within the cell is correspondingly shorter and the detecting efficiency measured based on the area under radioactivity curve must be lower. However, an increase in the volume causes a decrease in the resolution, since X and another radioactive component Y entering shortly after X emerge together in the cell.

It should be possible to obtain a high detecting efficiency without sacrificing the resolution, when a series of small counting cells of equal volume are connected longitudinally and the radioactivity signals detected by each counting cell are accumulated in synchronization with the traveling speed of the radioactive substances. Figure 2 illustrates the principle of SARD using three counting cells. A small fragment in a sample travels through the counting cells C_1 C_2 and C_3 with the lapse of time. The volume of each counting cell is equal to that of the counting cell I shown in Figure 1. Here, *a* is defined as sampling time. The idea is to accumulate the radioactivity signals toward the direction indicated by arrows into an accumulation counter.

A block diagram of SARD using three counting cells is shown in Figure 3. SARD is composed of three parts, a radiodetector assembly, switching units B and accumulation counters A. The radiodetector assembly comprises a series of small counting cells C of equal volume which are connected longitudinally. Each switching unit has one output terminal and three input terminals. A contact arm connects the output terminal with one of the input terminals. The working principle involved is that the radioactivity signals of a small fragment of the sample are accumulated into the same accumulation counter by means of the switching unit. When their detection by the third counting cell is completed, the accumulated signals are sent to a recorder, and the counting cycle is reset to the initial state. It is not necessary to ensure uniformity in the background counts and the counting efficiency of

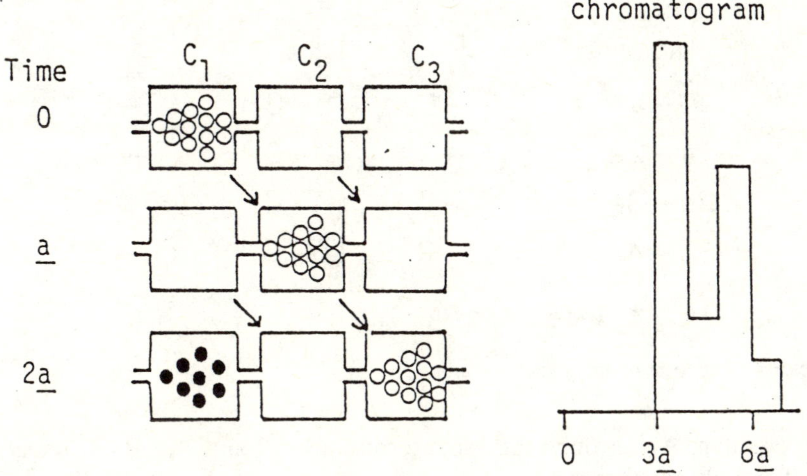

Figure 2. Principle of a synchronized accumulating radioisotope detector (SARD, n=3)

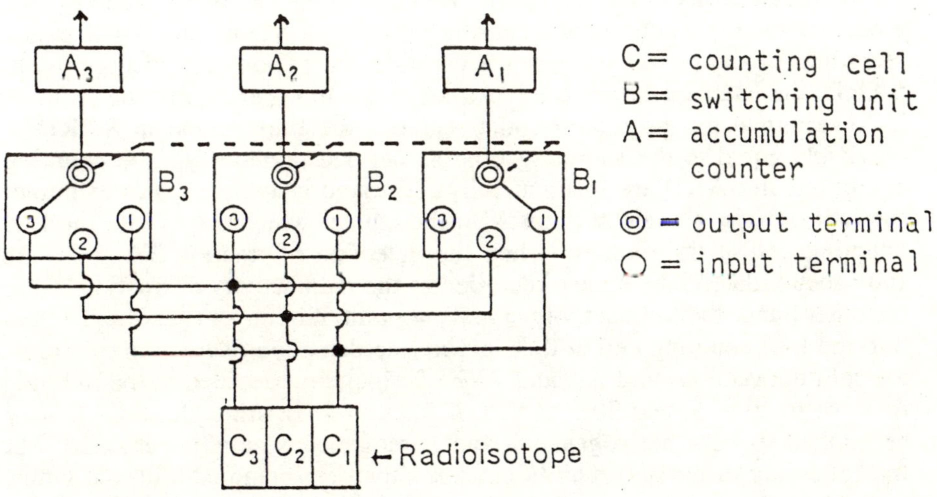

Figure 3. Block diagram of SARD (n=3)

each counting cell, since all portions are equally treated by all the counting cells.

Figure 4 explains the function of the switching unit. The movement of the contact arm in each switching unit can be synchronized with the flow rate of the radioactive substances. The radioactivity of the first portion is detected by C_1, C_2, and C_3 during the time intervals 0 - *a, a* - *2a and 2a*- *3a*, respectively, and the radioactivity signals are sent into the same accumula-

Time interval	A_1 – B_1	A_2 – B_2	A_3 – B_3
0 – a	① – C_1	② – C_2	③ – C_3*
a – 2a	② – C_2	③ – C_3*	① – C_1
2a – 3a	③ – C_3*	① – C_1	② – C_2
3a – 4a	① – C_1	② – C_2	③ – C_3*

* Reset time point

Figure 4. Function of the switching unit

tion counter, A_1, through the input terminals 1, 2 and 3 of B_1. The accumulated signals are then sent to a recorder and recorded as the radioactivity of fraction 1. The same procedures are followed for the next portion entering the detector assembly, the radioactivity signals being accumulated in A_3, and recorded as the radioactivity of fraction 2.

The radioactivity emerging from a HPLC column can be recorded by two methods. One is a continuous recording through a rate meter. Another is a recording of the radioactivity per a certain time period as a histogram. In SARD, the latter method is used. Although the time period may be arbitrarily determined by the conventional method, the time period in SARD is inevitably equal to the sampling time as defined previously. The counting conditions in SARD are schematically illustrated in Figure 5. For example, the radioactivity signals at point X in the eluate, which enters into the first counting cell at the moment when the detection of fraction 2 begins, are fully accumulated into A_3 and recorded as the radioactivity of fraction 2. On the other hand, the radioactivity signals at point Y in the eluate which enters into the first counting cell at 0.3a before the detection of fraction 5 begins, are split between A_3 and A_1, and 70% of which are recorded as the radioactivity of fraction 5, and 30% as that of fraction 4. In principle, SARD may be applied to both heterogeneous and homogeneous counting method, but the following discussion mainly concerns the data obtained with the latter.

EXPERIMENTAL

Instrument

A Shimadzu LC-2 high-performance liquid chromatograph equipped with a Shimadzu UVD-2 UV detector was used. A photograph of the radio-HPLC system is given in Figure 6. A block diagram of the radio-HPLC equipped with SARD of homogeneous counting type and the counting cell are shown in Figures 7 and 8, respectively. For a comparison of the conventional and the present method, the first counting cell was used not only as part of SARD but also as an independent conventional detector.

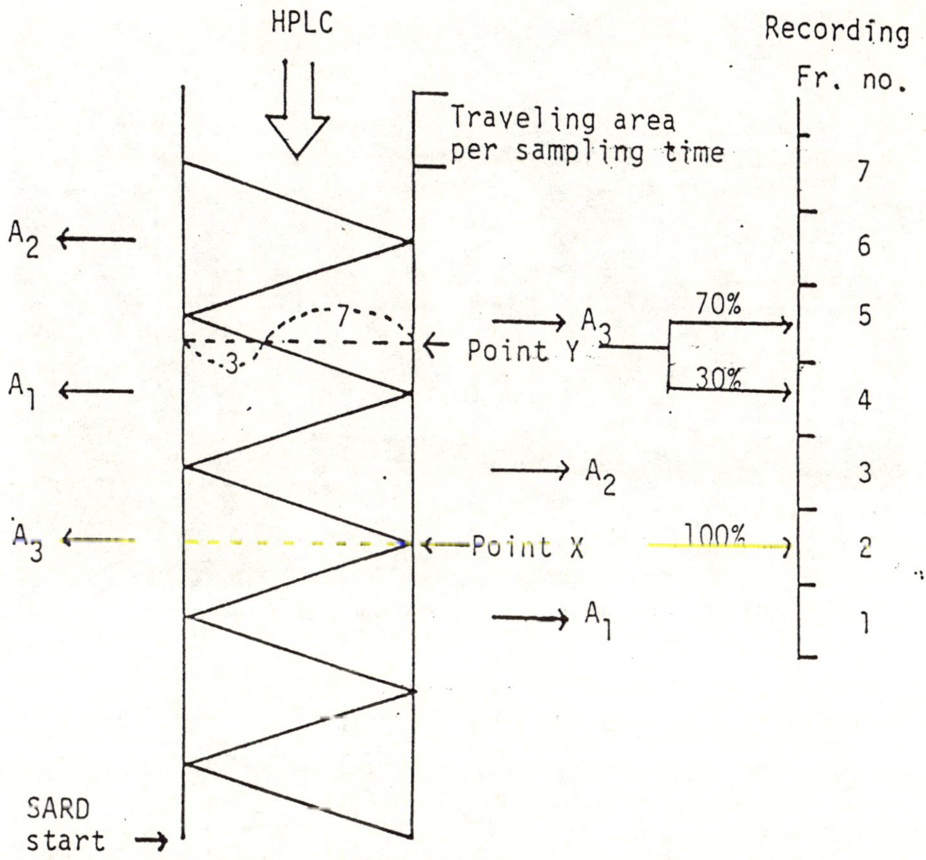

Figure 5. Counting conditions in SARD (n=3)

Operating conditions

Conditions for HPLC: For [8-^{14}C]adenine, Nucleosil 5, C_{18} column (4.6 × 150 mm), with a mobile phase of 0.01 M $HClO_4$-methanol (10:1), flow rate 0.5 ml/min. For [methylene-^{14}C]imipramine, Lichrosorb Si 60 column (4.6 x 150 mm), acetonitrile:methanol:ammonium water (25%) (100:20:1) mobile phase, flow rate 1.5 ml/min. In the homogeneous counting method, the eluate from HPLC was mixed with a liquid scintillator of hydrophylic type [DPO 4g, dimethyl POPOP 0.4g, naphthalene 100 g dissolved in dioxane:toluene:ethylcellosolve (750:150:100)] and the resulting solution (8.3 ml/min) was passed through the five counting cells having effective cell volume of 1.1 ml per cell. The sampling time was 8.0 seconds under these conditions. In the heterogeneous counting method, the eluate of [^{14}C]adenine was diluted with the mobile phase, and the resulting solution (1.5 ml/min) was led to the five counting cells packed with lithium-glass scintillator (effective volume: 0.14 ml). The eluate of [^{14}C]imipramine was

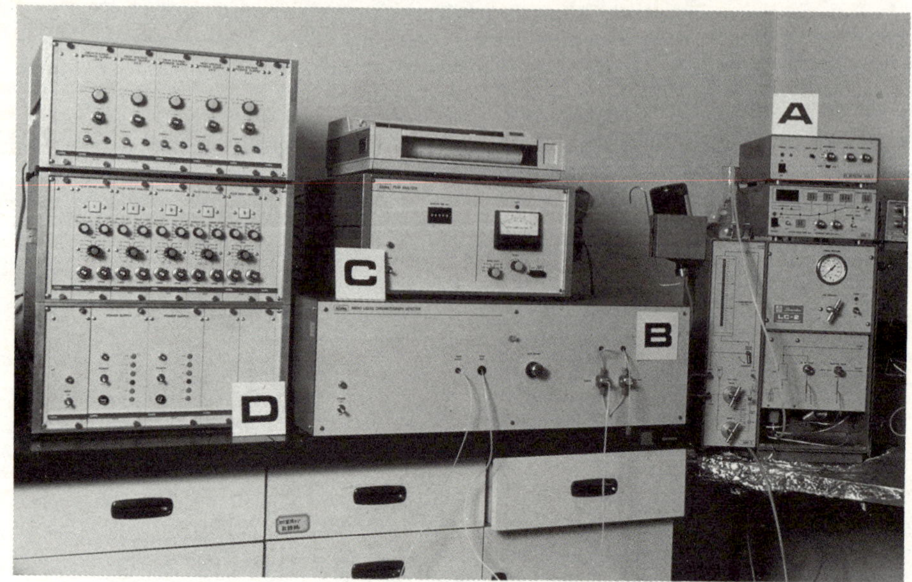

Figure 6. Photograph of the radio-HPLC equipped with SARD (n=5).

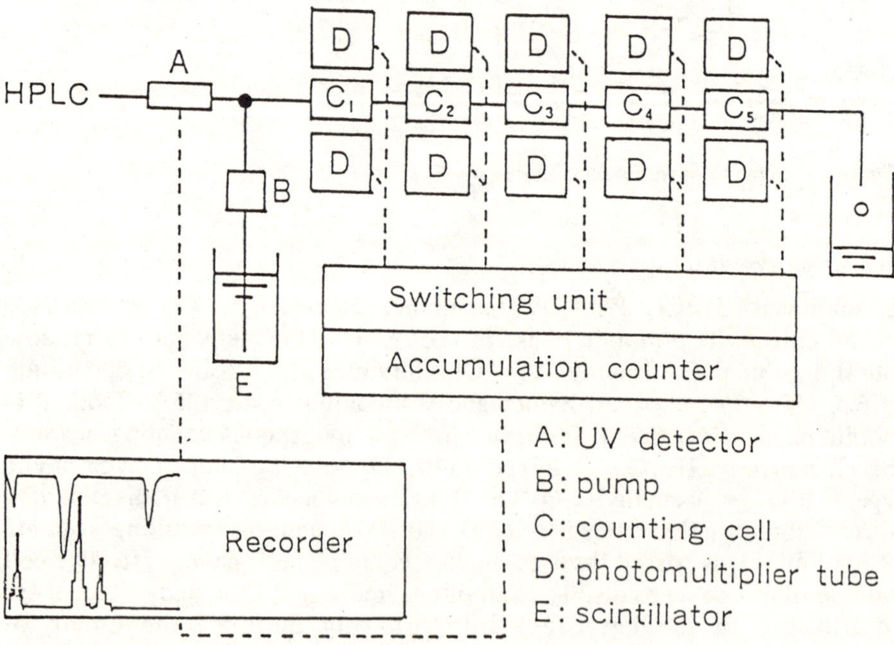

Figure 7. Radio-HPLC system equipped with SARD (n=5)

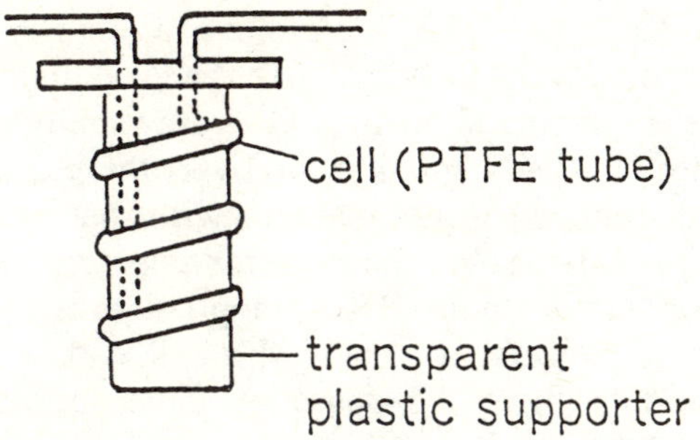

Figure 8. Homogeneous counting cell

led to the five counting cells packed with yttrium silicate glass. In the conventional method, the radioactivity was recorded with a time constant of 10 seconds.

Measurement of peak intensity

In SARD, simultaneously with the radioactivity recording, the fraction number, the counts per sampling time and the integral counts were typed out. The peak intensity was easily calculated by subtracting the integral counts of a fraction corresponding to just the beginning of the peak from that of the end of the peak. In the conventional method, the peak intensities were determined by measuring the peak area on the chromatograms.

RESULTS AND DISCUSSION

In the radio-HPLC equipped with SARD, the volume of each counting cell affects definitively the chromatographic resolution. With radioactivity monitors, it is necessary to compromise sensitivity versus resolution. Although it is important that the cell volume should be as small as possible to reduce band spreading, response of this detector is a function of the total amount of radioactivity in the counting cell at any time, and any reduction in the cell volume is accompanied by corresponding reduction in sensitivity. Therefore, in monitoring radioactivity a compromise between resolution and detector sensitivity must be reached, the exact nature of which depends on the analytical requirements (Snyder and Kirkland, 1979). Where the maximum resolution of HPLC is required, for example, in the analysis of a complex mixture, band-spreading arising from the detector cannot be tolerated, and this can only be avoided by sacrifices in the sensitivity of the radioactivity monitor. Alternatively, if only low resolution is required, the cell volume can be increased in order to enhance overall sensitivity and improve quantification (Reeve and Crozier, 1977). Reeve and Crozier recommended to choose a cell with a volume one third of that of the narrowest chromatographic peak of interest. Also, they mentioned that under these circumstances high sensitivities are obtained with no more than a 10% increase in peak width (Reeve and Crozier, 1983). We have attempted to develop the radio-HPLC equipped with SARD that might be used mainly in metabolism studies. The samples in the metabolism studies are usually of low radioactivity and their sizes are relatively small. Moreover, the number of radioactivity peaks are greatly reduced in comparison with that of mass peaks. For these reasons, it is reasonable that the detector sensitivity should be preferred to the resolution in the radio-HPLC applied to a metabolism study. From these theoretical aspects we designed a cell with 1.1 ml volume for homogeneous counting. Under the operating conditions described in Experimental, the transit time in the detector assembly is calculated to be 40 seconds.

The usefulness of SARD as a detector for radio-HPLC is largely dependent on how much peak broadening occurs due to the prolonged transit time of the radioactive substances in the multiple counting cells. In the case of the radio-GLC, no peak broadening was observed by the multiplying counting tubes. However, the multiplying counting cells for radio-HPLC may result in an appreciable peak broadening, since liquid suffers from back diffusion more extensively than gas. This was examined by comparing the chromatograms obtained by injecting 4.5 nCi (homogeneous counting) or 5.4 nCi (heterogeneous counting) of [^{14}C]imipramine and by monitoring the radioactivity simultaneously with the first and fifth counting cells of SARD (Fig. 9). In the homogeneous counting method, the peak broadening judged by the half width values, was not so large that SARD could not be applied to HPLC. This was proved by the comparison between the chromatogram A (the first counting cell) and B (SARD) in Figure 10. On the other hand, in the heterogeneous counting method peak broadening occurred to a significant extent, and means for reducing peak broadening will have to be devised when SARD is applied to the heterogenous counting method.

The advantages of SARD can be understood most clearly by comparing the radiochromatogram (Fig. 10) obtained simultaneously by the two operation modes. The sample is 0.54 nCi of [^{14}C]imipramine. The peak width of the chromatogram B (SARD) is comparable to that of A (the conventional). On the other hand, the integral intensity of B is approximately five times that of A. These experimental results suggest that the synchronized accumulation works also in SARD applied to HPLC, exactly as theoretically predicted. The reproducibility of measurements was compared between SARD and the conventional method by repeating the above mentioned experiment ten times. The experimental results are summarized in Table 1. The reproducibiltiy of SARD was improved by a factor of $\sqrt{5}$, since the transit time was extended five times.

The counting efficiency of this detector cell was estimated as follows: When 4.39 nCi of [^{14}C]adenine were injected, the average total radioactivity under the peak was 6182 counts. The transit time of the HPLC eluate in the detector assembly was calculated to be 0.667 minute from the volume of the detector assembly (1.1 × 5 ml) and the total flow rate (8.3 ml/min) of

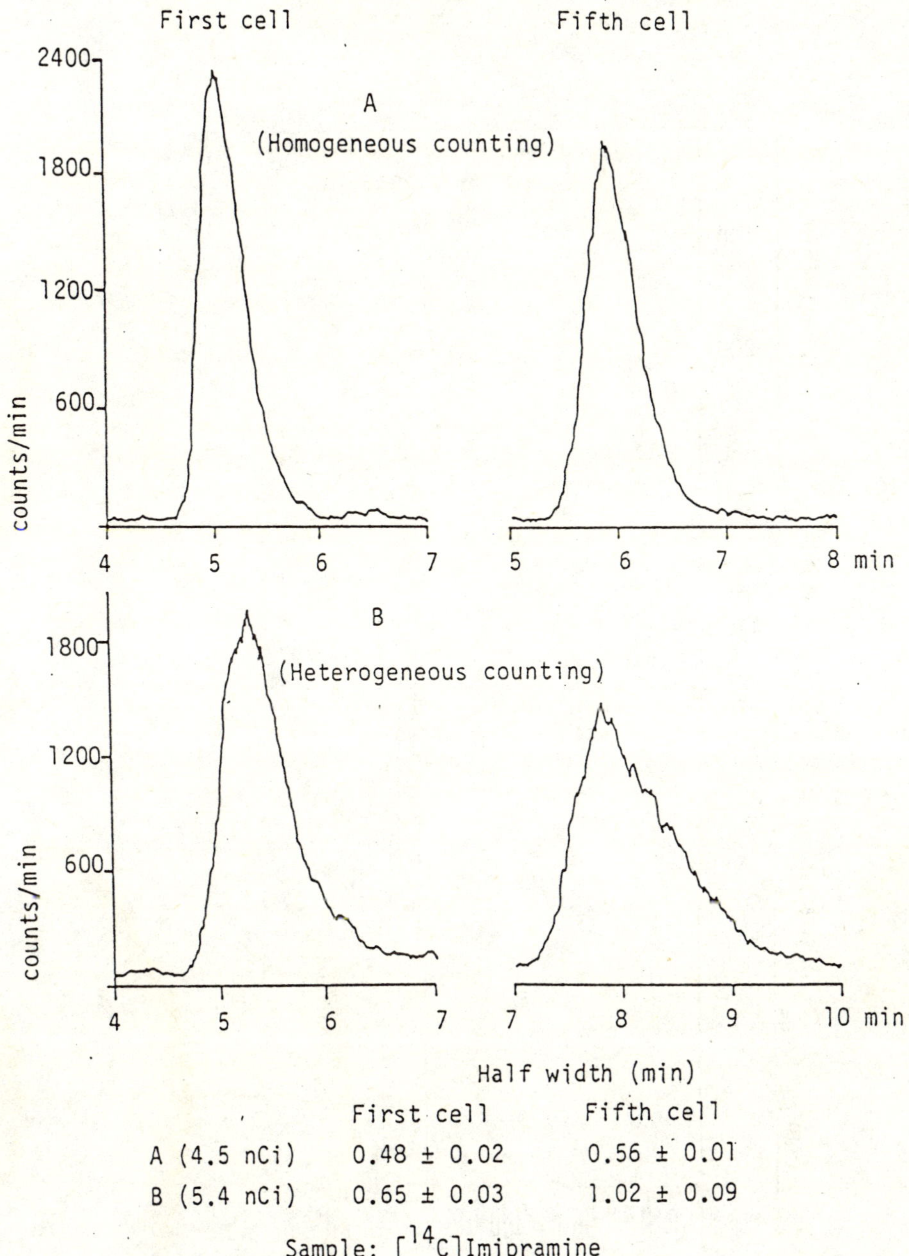

Figure 9. Peak broadening by the multiplying counting cells

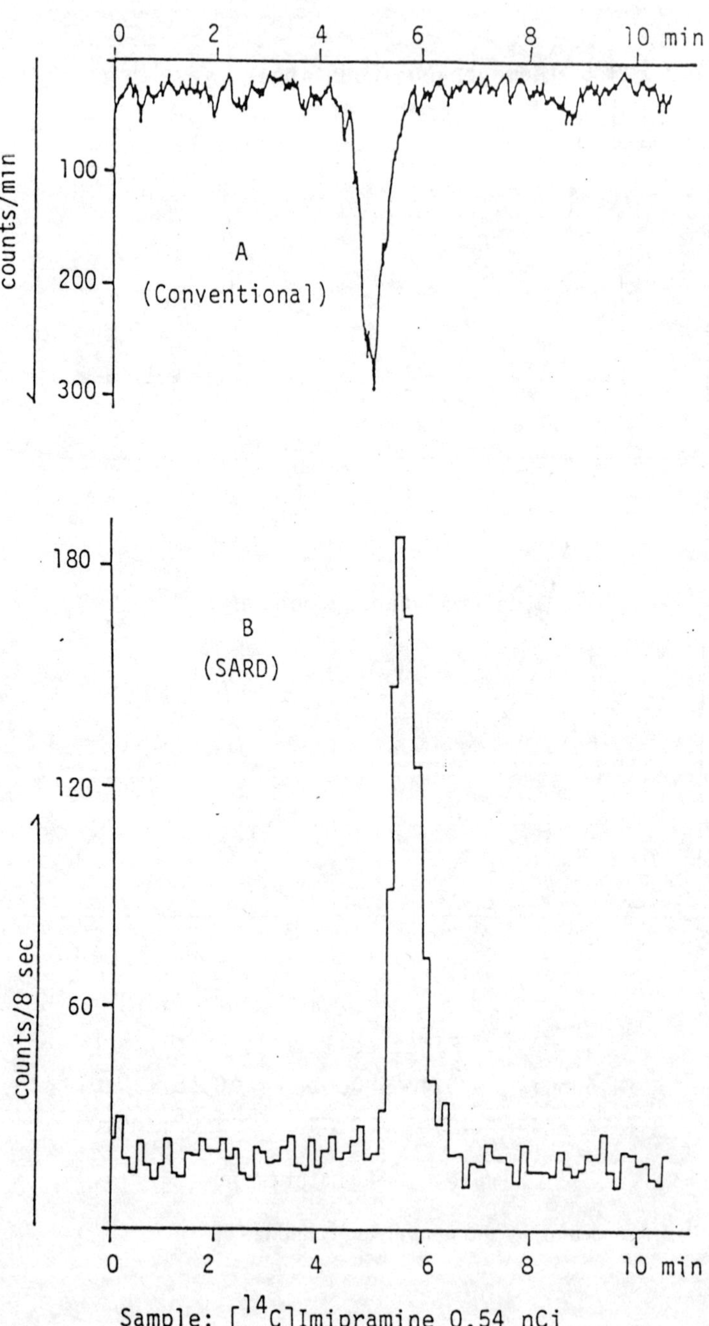

Figure 10. Comparison of two radiochromatograms

Table 1. Reproducibility of measurement (counts)*

	Conventional	SARD
Detector Volume (ml)	1.1	1.1 × 5
	125	663
	127	675
	132	658
	116	666
	114	659
	153	731
	130	682
	140	730
	122	694
	123	714
Mean	128.2	687.2
SD	11.5	28.6
CV (%)	9.0	4.2

*Sample: [^{14}C]imipramine 0.54 nCi

eluate and scintillator. From these data, the counting efficiency for [^{14}C]adenine in the homogeneous counting was determined to be 95.1%. In a similar manner, the counting efficiency for [^{14}C]imipramine was determined to be 82.9%. The counting efficiency for [^{3}H]adenine was calculated to be 34.6%. For more accurate determination of peak intensities, the fact that the composition of eluate may influence the counting efficiency slightly has to be taken into consideration. In the heterogeneous counting method, the counting efficiencies were 32.3% for [^{14}C]adenine and less than 0.1% for [^{3}H]adenine. The counting efficiency of the homogeneous counting cell was nearly the same as that of a liquid scintillation counter (LSC). This means that the radiochromatogram obtained by this radio-HPLC is equivalent to the radiochromatogram obtained by fractionating the HPLC eluate into 75 fractions per 10 minutes and counting each fraction for 40 seconds with LSC.

Linearities were examined by injecting a fixed radioactivity from 90 to 450 pCi [^{14}C]adenine into the chromatographic column. The experimental results are shown in Figure 11. The coefficients of correlation were 1.00 for SARD and 0.98 for the conventional method, demonstrating that the present method provides better quantitative accuracy.

The background counts were compared between SARD and the conventional method as shown in Figure 12. Although the detection limit of the conventional method is determined more or less subjectively, that of the present method may be treated as a matter of probability and estimated objectively from the transit time, the peak width, the standard deviations of background fractions, and the counting efficiency. For the conventional method the detection limit is usually discussed on the basis of signal-to-noise ratio (S/N) and S/N = 2 is usually used as the detection limit. How-

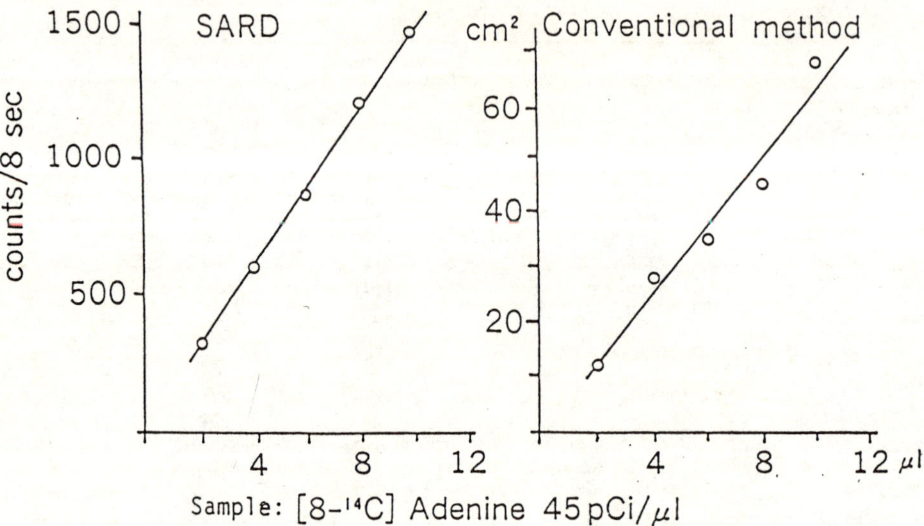

Figure 11. Comparison of linearity

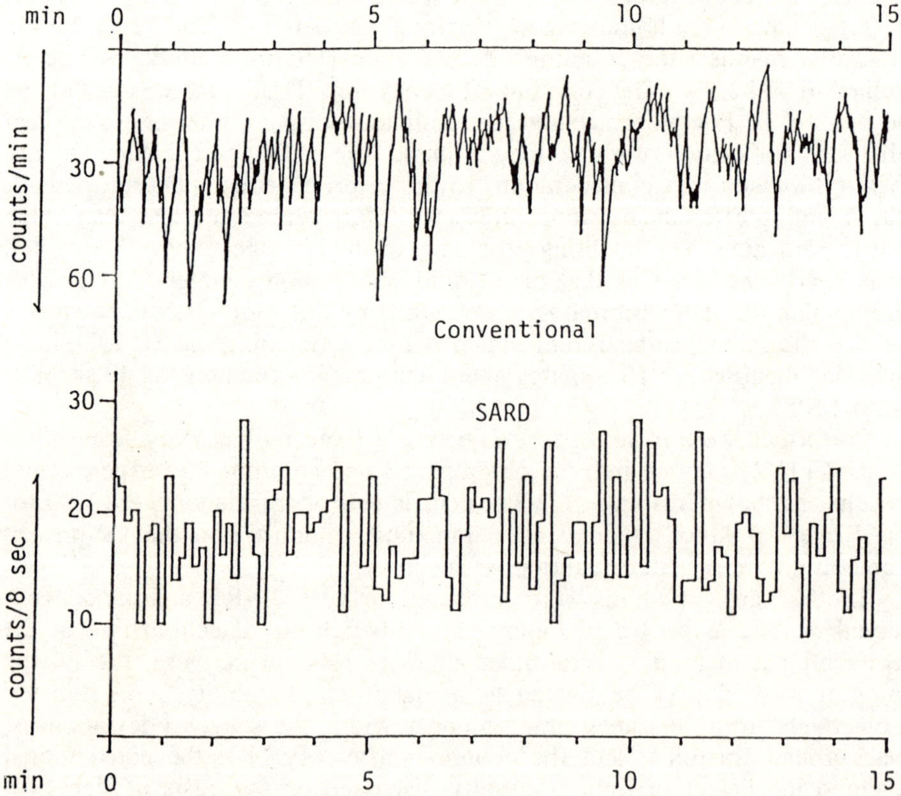

Figure 12. Comparison of background counts

ever, the detection limit of SARD is more reasonably discussed based on not the counts of one fraction but the total counts of several successive fractions, since the radioactivity in a peak appears over several fractions. We will propose the following parameters to define the detection limit of SARD.

$TCSF^{n}$: total counts of *n* successive fractions
$TCSF^{n,\ back}$: $TCSF^{n}$ in background region
$TCSF^{n,\ max}$: $TCSF^{n}$ which gives the highest value in the region of the peak of interest (corrected for background counts)

Here, *n* is the number of fractions which make $TCSF^{n,\ max}$ approximately 60% of the total counts of the peak of interest. The detection limit of SARD may be defined as:

$3 \times$ (SD of $TCSF^{n,\ back}$).

As an example, the detection limit [*x*(pCi)] of [^{14}C]imipramine was estimated as follows: The number of fractions that are taken into consideration should be determined depending on the spread of the peak of interest. The ratio of $TCSF^{3,\ max}$/total counts of the peak (see Fig. 10) was 0.615 ± 0.032. $TCSF^{3}$ was then used as a criterion for detection limit. In the case of [^{14}C]imipramine these parameters were as follows:

The transit time (T*t*) : 0.667 min
The counting efficiency (*E*) : 82.9%
SD of $TCSF^{3,\ back}$: 7.4 (see Fig. 12)

Therefore,

$$7.4 \times 3 = TCSF^{3,\ max} = x(\text{pCi}) \times 2.22 \times 0.667 \times 0.615$$
$$x = 29.4 \text{ pCi}$$

Since the radioactive disintegration suffers from statistical fluctuations, two probabilities should be considered in discussing the detection limit. One is the probability (α probability) that $TCSF^{n,\ back}$ exceeds the detection limit, giving a false peak. The other is the probability (β probability) that $TCSF^{n,\ max}$ of a sample containing a certain amount of radioactivity is below the detection limit, giving no peak. When $3 \times$ S D of $TCSF^{n,\ back}$ is the detection limit, the α probability is less than 0.14%. The detection limit calculated above is a value, where the β probability is 50%. The relationships between the amount of the radioactivity injected and the probability with which a peak is characterized is now under investigation. The detection limits for [^{14}C]imipramine were experimentally examined and one example is given in Figure 13. In this case, the detection limits of SARD and the conventional method were 36 and 180 pCi, respectively.

In the following, the detection limit of the radio-HPLC applied in a metabolism study is estimated. Supposing the molecular weight of the drug in question is 300 and the specific radioactivity is 30 mCi per mmol (which means the drug is labeled by one atom of carrier free ^{14}C and is diluted by an

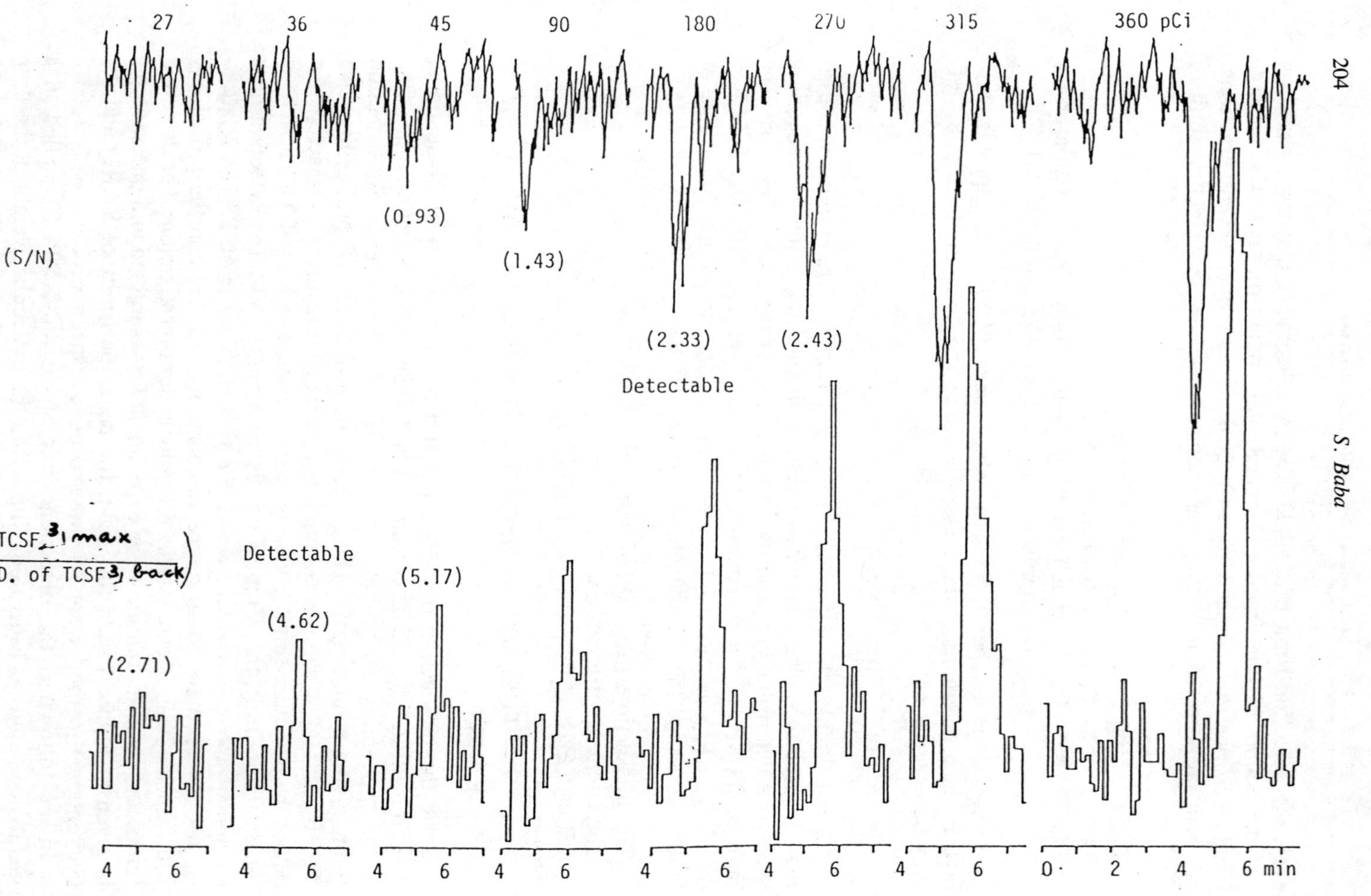

Figure 13. Detection limit

equal amount of carrier) and the parameters mentioned above are the same as imipramine, the detection limit is calculated to be approximately 0.3 ng per injection.

The resolution experiments were carried out by injecting 1.4 nCi of [^{14}C]imipramine at the time intervals of 60, 50, 40, 35, 30, 25, and 20 seconds. The experimental results are shown in Figure 14. The resolution of SARD was almost the same as that of the conventional method, but much lower than that of a UV detector.

APPLICATIONS

The radio-GLC and HPLC equipped with SARD offer two advantages. One is to provide improved detecting efficiency without sacrificing chromatographic resolution. The other is that the radioactivity information can be obtained in digital form. The latter advantage makes it possible to correlate easily the radioactivity information obtained with SARD with that obtained with LSC. The radio-HPLC has an additional advantage, in spite of the lower resolution, in comparison with GLC. Since the recovery of metabolites by HPLC is almost quantitative and the sensitivity for radioactivity is, in general, not influenced by chemical structures of metabolites, the amount of each metabolite can be directly estimated from the peak intensity (without preparation of calibration curve). By this radio-HPLC, it is possible to detect easily radioisotopically labeled metabolites among large amounts of endogenous metabolites, and to make a balance sheet of the amounts of the injected radioactivity and those of the radioactivity identified as metabolites on the radiochromatogram. This radio-HPLC can be used as a metabolic pattern analyzer. Furthermore, time courses of plasma levels of each metabolite may be clarified without sacrificing such small experimental animals as rats.

Figure 15 shows an example of the radio-HPLC applied to the clarification of species differences in the metabolism of a drug. The samples were obtained from the urine of [^{14}C]suprofen administered experimental animals (Mori *et al.*, 1983). The amounts of radioactivity administered was less than 4 nCi. The study was successful with a relatively small amount of radioactivity and without tedious procedures such as fractionation followed by liquid scintillation counting.

The next example is an analysis of glyceryl trinitrate (GTN) and its metabolites in rat plasma (Baba *et al.*, 1984). Three male rats, weighing approximately 300 g, received a single intravenous dose of [^{14}C]GTN (sp. radioactivity: 224.6 μCi/mg, 112.3 μCi/kg) dissolved in 0.3 ml of saline. Heparinized blood samples (0.15 - 0.20 ml) were taken. The plasma samples were divided into two sets. Each plasma sample in one set (0.02 ml) was subjected to LSC and the total radioactivity in the plasma was determined. Each plasma sample in the other set (0.05 ml) was subjected to radio-HPLC analysis. The estimation of the amounts of [^{14}C]GTN and its metabolites in plasma was based on the relative peak intensities on the

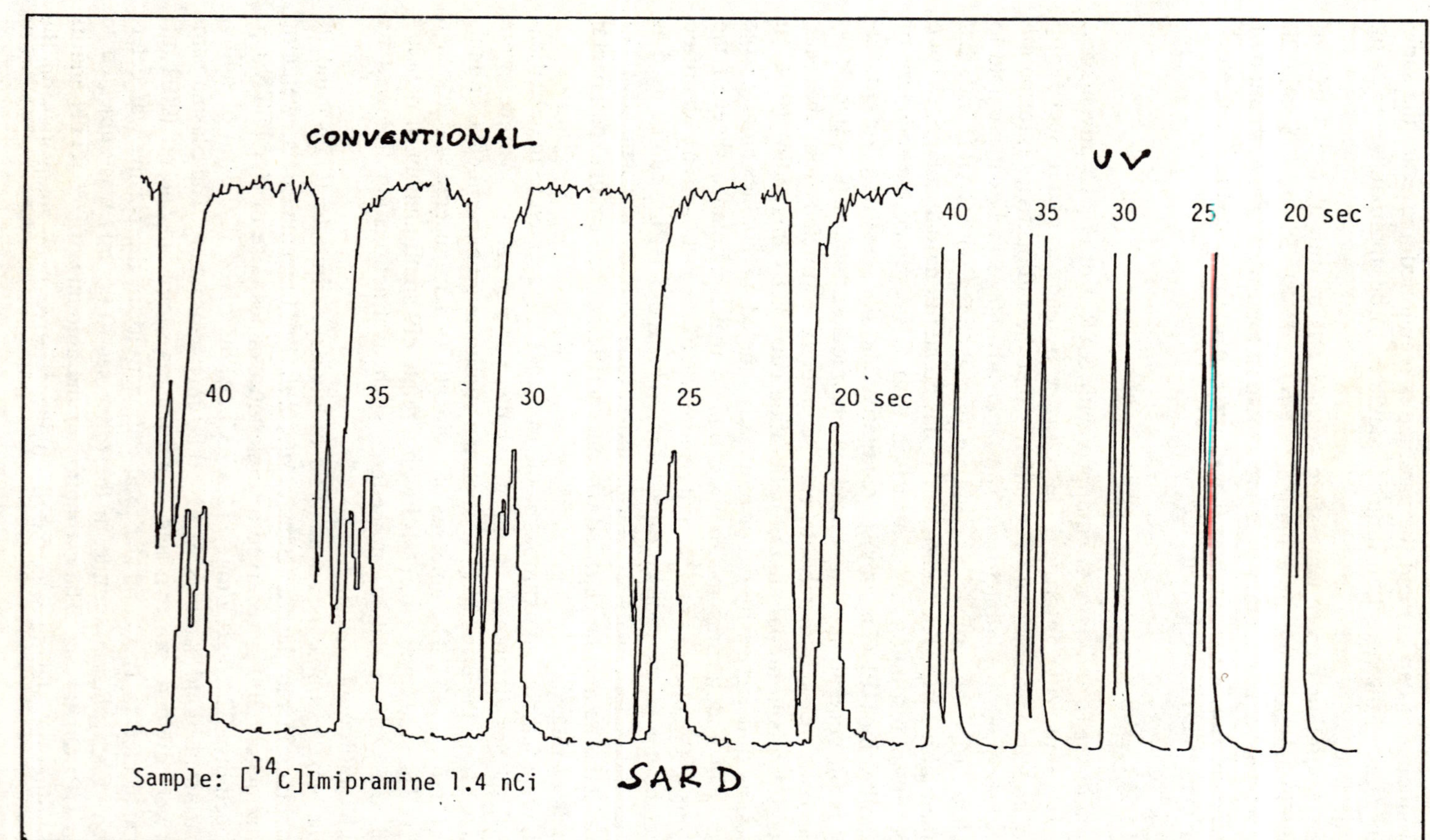

Figure 14. Comparison of resolution

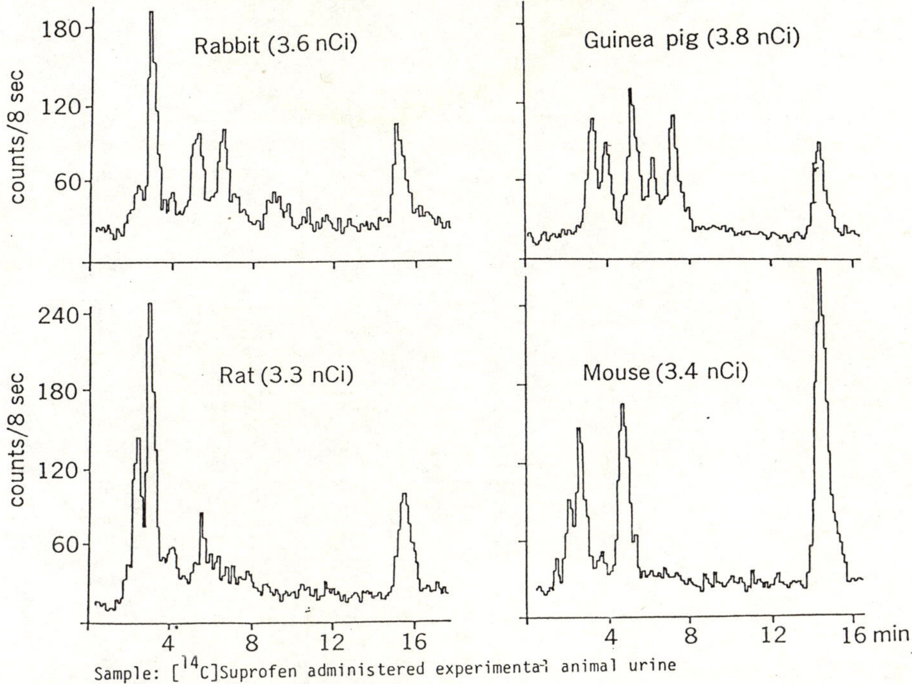

Figure 15. Metabolic pattern analysis

chromatogram and the total radioactivity measured by LSC. An example of the radiochromatograms thus obtained is shown in Figure 16.

The reproducibility in counting of the radio-HPLC was determined by injecting the same test plasma sample into the radio-HPLC system four times. The results are shown in Table 2. The present method obviously provides very good reproducibility of counting. The within-day precision of the radio-HPLC analysis for [^{14}C]GTN and its metabolites was good, as assessed by conducting replicate (n = 4) analyses of the plasma sample taken at five minutes after intravenous administration of [^{14}C]GTN.

One example of the time courses of plasma concentration of [^{14}C]GTN and its metabolites following intravenous administration of [^{14}C]GTN to a rat is shown in Figure 17. The half-lives of [^{14}C]GTN and its metabolites are presented in Table 3. The data give some insight into interindividual differences. It should be pointed out that the use of the radio-HPLC, which gives satisfactory resolution with high detection efficiency, is of great advantage to evaluate interindividual differences of pharmacokinetics in small laboratory animals.

The time course of plasma levels of a drug is often discussed on the basis of total radioactivity in plasma after the administration of the radioisotopically labeled drug. For detailed pharmacokinetics studies of a drug it is necessary to know the time course of not only the substrate but also its

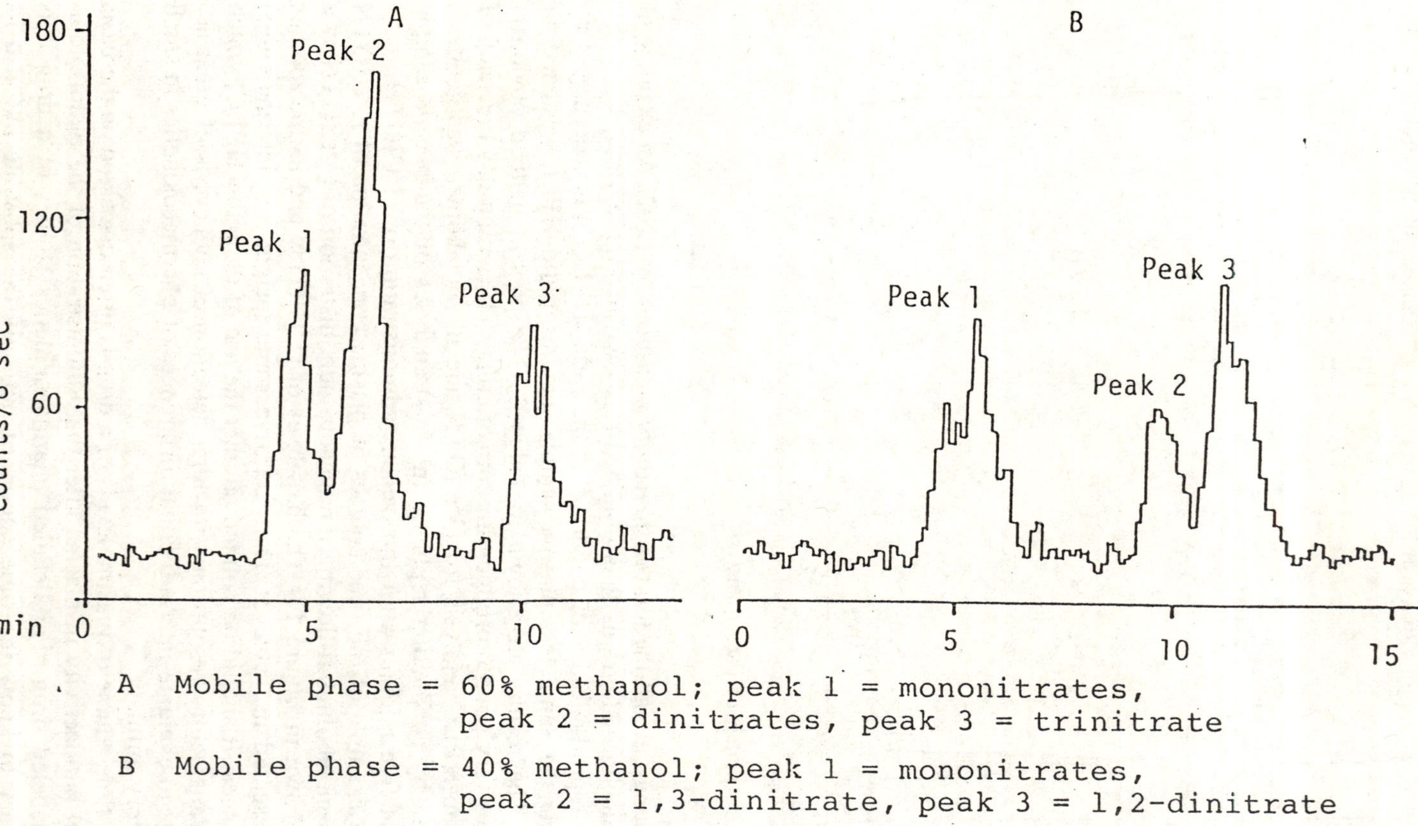

Figure 16. Radio-HPLC of an extract of plasma sample at 2 minutes after intravenous administration of [^{14}C]glyceryl trinitrate (112.3 μCi/kg)

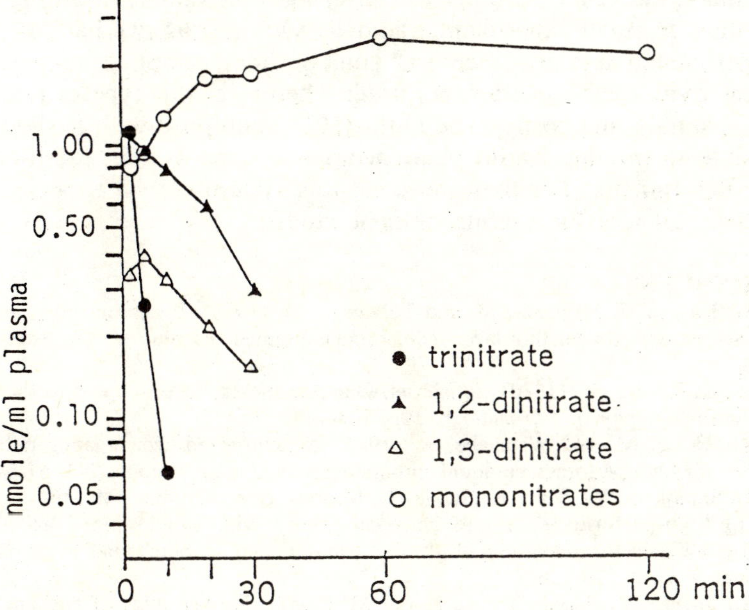

Figure 17. Plasma concentrations of glyceryl trinitrate and its metabolites in rat

Table 2. Reproducibility of analyses of [^{14}C]glyceryl trinitrate and its metabolites in rat plasma

	ng/ml plasma		
	mononitrates	dinitrates	trinitrate
1	42.7	118.7	27.3
2	43.7	115.7	27.7
3	42.3	116.7	28.3
4	41.7	116.3	24.7
Mean	42.6	116.9	27.0
SD	0.84	1.30	1.59
CV (%)	2.0	1.1	5.9

Table 3. Plasma half-lives (min) of [^{14}C]glyceryl trinitrate and its metabolites in plasma after intravenous administration of [^{14}C]glyceryl trinitrate in male rats

	1.2-dinitrate	1,3-dinitrate	trinitrate
1	14.6	21.2	1.9
2	15.3	9.1	2.6
3	14.3	11.1	2.1
Mean	14.7	13.8	2.2
SE	0.29	3.75	0.21

metabolites. However, in order to clarify the time course of a drug and its metabolites in small experimental animals such as rats, we had to sacrifice the experimental animal at each time point of blood sampling, because of the low sensitivity of the instruments used. Therefore, this type of research is time consuming and costly. The radio-HPLC equipped with SARD makes it possible to provide useful pharmacokinetic data without sacrificing experimental animals. For these reasons, this system can be expected to become a useful tool for pharmacokinetic studies.

REFERENCES

Baba, S., Kasuya, Y., Takeda, M. and Tokunaga, N. (1979). Synchronized accumulating radioisotope detector for thin-layer radiochromatographic scanning. *J. Chromatogr.* 168, 49-58.

Baba, S. and Kasuya, Y. (1980). Synchronized accumulating radioisotope detector for radio gas chromatography. *J. Chromatogr.* 196, 144-149.

Baba, S., Horie, M. and Watanabe, K. (1982). Synchronized accumulating radioisotope detector for high-performance liquid chromatography. *J. Chromatogr.* 244, 57-64.

Baba, S., Shinohara, Y., Sano, H., Inoue, S., Masuda, S. and Kurono, M. (1984). Application of high-performance liquid chromatography with synchronized accumulating radioisotope detector to analysis of glyceryl trinitrate and its metabolites in rat plasma. *J. Chromatogr.* 305,119-126.

Baba, S., Suzuki, Y., Sasaki Y. and Horie, M. (1987). Further study of the synchronized accumulating radioisotope detector for high-performance liquid chromatography. *J. Chromatogr.* 392, 157-164.

Mori, Y., Yokoya, F., Toyoshi, K., Baba, S. and Sakai, Y. (1983). Absorption and excretion of suprofen in rats. *Drug Metab. Dispos.*, 11, 387- 391.

Reeve, D.R. and Crozier, A. (1977). Radioactivity monitor for high-performance liquid chromatography. *J. Chromatogr.* 137, 271-282.

Reeve, D.R. and Crozier, A. (1983). HPLC radioactivity monitors-fact and fiction. *Lab. Practice,* 32, 59-60.

Snyder, L.R. and Kirkland, J.J. (1979). *Introduction to Modern Liquid Chromatography.* John Wiley & Sons, Inc. New York, Chichester, Brisbane, Toronto, pp. 158-161.

Subject Index